The Application
of Regression Analysis

The Application of Regression Analysis

Dick R. Wittink

Cornell University

Allyn and Bacon, Inc.
Boston   London   Sydney   Toronto

To the 3 M's: Marian, Marsha, and Mark

Copyright © 1988 by Allyn and Bacon, Inc.
A Division of Simon & Schuster
160 Gould Street
Needham Heights, Massachusetts 02194-2310

Series Editor: Rich Carle, Cary Tengler
Production Administrator: Annette Joseph
Production Coordinator: Susan Freese
Editorial-Production Service: Publication Services
Cover Administrator: Linda K. Dickinson
Cover Designer: Lynda Fishbourne

Library of Congress Cataloging-in-Publication Data

Wittink, Dick R.
    The application of regression analysis.

    Includes index.
    1. Regression analysis. I. Title.
QA278.2.W58   1988   519.5'36   87–30740
ISBN 0–205–11252–8

Printed in the United States of America
10 9 8 7 6 5 4 3 2 1   92 91 90 89 88 87

Contents

Preface

Regression analysis is probably the most common statistical tool used in business applications as well as in academic projects. Its popularity has grown because of the speed with which computers can provide regression results, the availability of data, and the quality of software. In most business and social science curricula, students are required to learn at least the fundamentals of regression analysis.

In practice, users of regression analysis often feel they lack sufficient exposure to be able to carry out a project with confidence. Some users appear to have confidence in their results, only to be disappointed when the decisions based on the results turn out to deviate strongly from the expected outcomes. It is likely that the tool of regression analysis is misused far more than it is used properly.

There are many possible reasons for such misuse of regression analysis. For example, the strong emphasis on the statistical aspects may lead regression users to believe that it is a simple procedure to produce a useful and valid application. Algorithms such as stepwise regression have been developed to ease the task of users. However, the availability of algorithms should not influence regression users to ignore the importance of up-front thinking to guide the initial model specification. Perhaps the most critical components to the development of a valid result are the identification of appropriate variables and the manner in which variables are allowed to relate to each other.

In this book, we provide a broader perspective on regression analysis than is contained in most texts. We emphasize that substantive knowledge of the application is required. This knowledge must be used prior to any data analysis. The estimates obtained should be investigated to see if the implications are reasonable, and the unexplained variation must be subjected to tests. In addition, we propose a framework that can be used to evaluate regression results obtained by others as well as to produce such results.

This book can be used in any course in which a considerable amount of time is spent on regression analysis. It is possible to pick and choose selected aspects in this book to give students a flavor of the issues. Having the book makes it less likely for regression users to ignore the framework proposed in the text. Of course, the more regression users are exposed to all the material, the more likely it is that the ultimate applications will be successful ones.

Our experience is that students benefit from assignments that force them to focus on the identification of a problem for which regression analysis is applicable. In such assignments, students can be asked to specify relevant variables, to justify the variables selected, and to indicate, at least graphically, the nature of the likely relation between variables. If a sufficiently large part of the course involves regression analysis, we also require students to carry out a project on their own. For this project they must complete all the aspects identified in the proposed framework. Such hands-on experience gives students the opportunity to put the material to use and to develop a better understanding of the difficulties associated with regression analysis in practice.

Acknowledgments

Over the years that I have taught regression analysis at Stanford and Cornell, many students have read my course notes. Their feedback and questions have stimulated me to expand the notes. They encouraged me to prepare a textbook because of their disappointment with available books. Without their stimulation this book would not have been completed. Two students have been especially helpful. Doug Bates provided constructive and detailed substantive and editorial suggestions on early drafts of selected chapters; and Pete Huff examined all chapters carefully and made numerous comments that were the basis for substantial improvements.

Several colleagues suggested useful and diverse perspectives regarding the chapter on R^2 (Chapter 8). The content of this chapter is perhaps the most controversial. Even though I may not agree with all their ideas, I have benefited from the comments made by Albert Bemmoar, Jay Russo, and Wilfried Vanhonacker.

Of course all the hard work was accomplished by someone else. Cindy LeFever was extremely kind and careful in putting everything on a word processor.

Finally, I wish to thank Allyn and Bacon's reviewers—Jeffrey E. Jarrett (University of Rhode Island) and Ronald Koot (Penn State University)—for their help with this project.

The Application
of Regression Analysis

1

Introduction to Regression Analysis

Regression analysis is an approach that may be used for the study of relations between variables—particularly for the purpose of understanding *how* one variable depends on one or more other variables. In this chapter, we define the term *regression*, introduce regression analysis, and describe two purposes of regression analysis.

Regression Defined

The term *regression* is often used to indicate "the return to a mean or average value." About one hundred years ago, Francis Galton published a paper in which he reported that the average height of sons with tall fathers is less than the fathers' height (both measured at adult ages). Similarly, the average height of sons with short fathers was reported to be greater than their fathers' height. Galton emphasized the "regression toward the mean" phenomenon in his data. He also found a positive relationship between the height of fathers and the height of their sons. Today, any study of relations between variables is often accomplished through regression analysis.

To understand Galton's result in the context of predictions and relationships, assume that the average height of males does not increase from one generation to the next. Specifically, let the average height of all adult males be 5 feet 6 inches or 66 inches. And suppose that we observe the data in Table 1.1.

Table 1.1
The average height of sons and their fathers' actual height

Sons' average height (inches)	Fathers' actual height (inches)
64.5	63
65	64
65.5	65
66	66
66.5	67
67	68
67.5	69

These data can be interpreted as follows. For all fathers with an actual height of 65 inches, the *average* height of their sons equals 65.5 inches, or for one father who has a height of 65 inches, we *predict* that his son will reach a height of 65.5 inches.

Now consider the relationship between the two variables measuring the height of fathers and the height of their sons. If a father is one inch shorter than 66 inches, we predict the height of his son to be half an inch shorter than 66 inches. This is an example of the relationship using one pair of the data (65.5, 65) from this table. Formally, based on all the data, we can describe the relationship as follows:

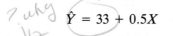

$$\hat{Y} = 33 + 0.5X$$

where $\hat{Y}$ is the predicted height of a son, in inches, and
X is the actual height of the son's father, in inches.

This equation *quantifies* the relationship between the fathers' actual height and their sons' average height. It provides a perfect fit for the data in Table 1.1. (For example, using $X = 68$ in the equation gives $\hat{Y} = 67$, which is exactly the number in the table.) However, the equation is not exact at the individual level. That is, a given son's height tends to deviate from the average (not shown in the table).

Although Galton used regression analysis to emphasize the "regression toward the mean" phenomenon, the term *regression analysis* is now used to refer to studies of relations between variables. The technique is used heavily in the social sciences, especially in economics and related disciplines. Regression analysis, then, is a technique for quantifying the relationship between a *criterion variable* (also called a dependent variable) and one or more *predictor variables* (or independent variables). This technique may be used for two main purposes: (1) to predict the criterion variable based on specified values for the predictor variable(s) and (2) to understand how the predictor variable(s) influence or relate to the criterion variable.

In our description of the technique, we emphasize the objective of *understanding*. That is, our primary objective is to discover and describe relations between variables as precisely and accurately as possible. Such understanding can be useful to people interested in extending or testing theories but also to people who want to improve the quality of their decisions. For example, a director of admissions at a graduate school of business may admit candidates partly based on the predicted academic performance. Knowing the relationship between academic performance (i.e., grade-point average) for the enrolled students and their credentials at the time of application (i.e., test scores, undergraduate grades and area of study, and work experience) may improve the director's decisions. Or a brand manager may use data on the brand's sales along with information about price, advertising support, and promotional programs to understand the influence of variables under the manager's control. Regression analysis has many applications in all areas of business and government, as well as other areas where a precise and valid understanding of relationships provides tremendous value. In particular, the quality of decisions often depends on the quantification of relationships between variables.

Value of Regression Analysis

Consider the admissions example described above as a possible application of regression analysis. In the case of an admissions director at a business school, the director's objective may be to admit the most highly qualified class in a given year. For example, the objective may be to admit applicants with appropriate academic qualifications and high career potential. In addition, one may specify that the class should be balanced (i.e., it should have diversity in work experience, career plans, and other factors). At least one of these objectives (admitted applicants should have high academic qualifications) can be satisfied by developing a regression equation that allows the prediction of academic performance for each applicant, in case the applicant is admitted. If it is possible to have an equation that provides accurate predictions, the director has more time to pursue other objectives (e.g., high career potential, diversity) that are more difficult to quantify.

Other benefits also result from the use of a regression equation to predict the academic performance of applicants. For example, an equation is perfectly consistent. Given specified values on the variables used to predict academic performance, the equation will always provide the same prediction. Human beings, on the other hand, are inconsistent. A given applicant may be judged as highly qualified if he or she is considered immediately after a set of applicants with relatively poor qualifications. The admissions officer's evaluation may be less positive if the same applicant is considered immediately after a set of applicants with impressive credentials.

In addition to the benefits of (1) allowing an admissions director to concentrate efforts on the more difficult aspects of the job and (2) providing consistent predictions, a considerable body of evidence suggests that regression analysis provides better predictions than decision makers can make without analysis. This superiority has been demonstrated for predictions of academic performance. Although this third benefit is not guaranteed for all regression applications, it is clear that regression analysis is a tool that (1) suggests (and quantifies) the nature of relations between variables, (2) provides consistent predictions, (3) may provide superior predictions, and (4) may save time (or allow a decision maker to focus more time and energy on nonquantifiable aspects).

Obtaining a Regression Equation

We have described the possible benefits of a regression equation for a director of admissions at a business school. Similar arguments apply to the decisions made by a brand manager. The manager may spend a considerable amount of time examining historical data and thinking about the possible influence of programs under his or her control, and those programs under competitors' control, on the brand's sales volume in a given period. Conclusions reached would influence the programs (e.g., advertising, promotions, temporary discounts) used in current and future periods. The manager could also use regression analysis to estimate the influence of such programs on the brand's sales.

One reason for using regression analysis is to have a systematic way of disentangling the influence of these marketing programs. It is extremely difficult for humans to determine the separate effects of different variables on a brand's sales. Hence, regression analysis may provide similar benefits to a brand manager as those identified for an admissions officer.

This discussion suggests that a decision maker may receive important benefits resulting from the use of an equation obtained from regression analysis. The *process* of obtaining this equation is, however, quite complex. Increasingly, managers in both the private and public sectors use regression analysis for a wide variety of problems. As the applications of regression analysis increase in number, so too does the extent of misuse of the technique. Today, data are abundantly available, and software packages for regression analysis are within reach of every computer user. Of course misuse is particularly likely from users who have only superficial knowledge of the technique and the assumptions underlying it.

The fact that misuse occurs does not imply that we should abandon the technique; rather, the technique should not be used unless we understand its assumptions and limitations. The technique can be used properly only after we have studied it in considerable detail, including the role of the assumptions, the ways in which possible violations of the assumptions can be examined, the implications if one or more assumptions is violated, and the remedies available in case of such problems. Our objective is to identify both the technique's potential and its limitations, using examples where possible.

2

Simple Regression Analysis Described

In this chapter, we discuss how regression analysis can be used to describe a relationship. In the next chapter, we discuss statistical inference (hypothesis testing, confidence intervals) in the context of regression analysis. In both chapters we confine ourselves to only two variables: a criterion variable and one predictor variable. This is called *simple regression analysis*. In later chapters we introduce additional predictor variables to describe multiple regression analysis, discuss how regression's assumptions can be checked, and present important extensions of the technique.

Predictions

Consider first the idea of making predictions about the value of a criterion variable. Many management-related problems involve such predictions or forecasts. For example, we may be considering an investment in a commercial project. The attractiveness of the investment should depend on, among other things, the return on the investment; however, actual return is usually quite uncertain. The investment must then be assessed on the predicted return on investment. Of course for this prediction to be useful, we would want a high level of accuracy. After many years of experience people may have certain skills and knowledge that assist them in making such predictions. Intimate knowledge of the factors relevant to a given situation and experience in a given line of work allow many people to make predictions with considerable accuracy. Thus, a manager faced with an investment decision may be able to rely quite successfully on his or her experience to determine whether and how much to invest.

On the other hand, if data have been gathered systematically on the returns achieved for several (many) investment opportunities, it may be possible to learn how the return depends on characteristics of the investment opportunities. In that case, we can use regression analysis to examine the relationship between the return and the characteristics of the investments. This does not mean that the manager's judgment is irrelevant. The manager's experience, knowledge, and background are necessary for specifying the variables that should be examined to predict the criterion variable. Regression analysis can only quantify the relationship between the variables identified as potentially relevant by a knowledgeable person.

Example 2.1
_____

Suppose you are in your last year of graduate studies or college. You are about to get a job offer, but you do not know what your annual starting salary will be. One way to approach this problem is to ask fellow students who have already received job offers and who know their starting salaries. Consider the following data obtained from six fellow students who each received one job offer:

Annual starting salary (dollars)

$$
\begin{array}{r}
20,000 \\
24,500 \\
23,000 \\
25,000 \\
20,000 \\
\underline{22,500} \\
\text{Total } 135,000
\end{array}
$$

Based on this sample of six observations, we can compute a mean or median and use this measure of central tendency to predict a starting salary.[1] In essence, we can use measures of central tendency and variability to summarize the data as follows. Let Y_i be the starting annual salary for the ith individual. The sample mean, $\bar{Y}$, is computed as

$$\bar{Y} = \frac{\sum\limits_{i=1}^{n} Y_i}{n} = \frac{\$135,000}{6} = \$22,500$$

If you feel that your earnings potential is similar to that of the other students, your starting salary prediction would be $22,500.

We can quantify the uncertainty of this prediction by measuring the variability in the starting salaries. The total amount of variation in the sample can be captured with the following computations.

Y_i	$(Y_i - \bar{Y})$	$(Y_i - \bar{Y})^2$
20,000	−2,500	6,250,000
24,500	2,000	4,000,000
23,000	500	250,000
25,000	2,500	6,250,000
20,000	−2,500	6,250,000
22,500	0	0
		23,000,000

Total amount of variation $\sum\limits_{i=1}^{n} (Y_i - \bar{Y})^2 = 23,000,000.$

The total amount of variation quantifies the variation (in squared dollars) in the sample. This quantity can be used in formulas for the variance or the standard deviation as an index of uncertainty about the prediction.

Although this simple computation of the mean starting salary could be satisfactory for some, we may believe that we could be more accurate. For example, there could be differences among the six starting salaries based upon the types of industry the offers came from or on the individual characteristics of the six students, such as their academic performances. In that case, a more accurate prediction can perhaps be made, if it is made conditional upon these student characteristics.

To keep it simple, suppose you believe that starting salary depends on academic performance. To investigate the possibility of such a relationship, you have asked the six fellow students about their grade-point averages. Assume that each individual reports truthfully his or her grade-point average, so that you now have six pairs of data on the two variables as shown:

Annual starting salary (dollars)	Grade-point average
20,000	2.8
24,500	3.4
23,000	3.2
25,000	3.8
20,000	3.2
22,500	3.4

To examine whether a relationship exists, and what the nature of that relationship may be, it is useful to draw a scattergram of the data, as shown in Figure 2.1. Even though we do not have many data points, we can identify a pattern in the data. There is a tendency for a positive relation to exist between the two variables: as the grade point average increases, the annual starting salary also tends to increase. Alternatively, we can say that students with higher grade-point averages tend to have higher starting salaries.

To quantify the relationship, we need to specify its functional form. The most convenient and common functional form is that of a linear relation between the two variables; however, this is not appropriate for all problems. Other relationships exist and will be described in Chapter 6. We must always ask ourselves whether it makes sense for a linear relationship to exist. As long as we are working with only two variables, we should *at a minimum* examine a scattergram of the data to see whether the assumption of a linear relationship is appropriate.[2] Based on the scattergram in Figure 2.1, we conclude that a linear function is a reasonable approximation of the relationship between the variables; however, we must remember that the number of data points is very small, so we cannot make this statement with a great deal of confidence. Also, if there are other relevant predictor variables to consider (such as the type of industry), this scattergram may provide invalid information (see Chapter 7).

We are often able, based on knowledge of the problem, to specify the nature of a possible relationship. Analysis of the data (for example, with the aid of a scattergram) is useful for checking whether the proposed linearity or nonlinearity exists in the sample. If

Figure 2.1
Scattergram of data on starting salary and grade-point average

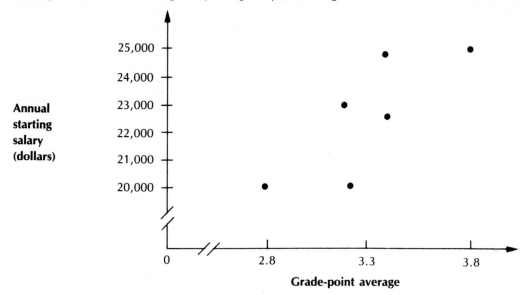

we have good reason to believe that the relationship is nonlinear, we should attempt to specify a functional form that accommodates this nonlinearity (see also Chapter 6).

Problems

For each data set given below, (1) construct a scattergram, and (2) argue the applicability of a linear function for the relationship between the two variables, given the data in the scattergram.

2.1

List price of selected homes (thousands of dollars)	Number of square feet (thousands of feet)
140	1.8
80	1.4
150	1.7
210	2.2
100	1.5
170	1.8
160	2.0
200	2.3

2.2	Demand for product M (thousands of units)	Price for product M (dollars)
	55	1.20
	54	1.15
	46	1.25
	36	1.45
	22	1.80
	20	1.70
	60	1.10
	42	1.50
	50	1.30
	30	1.60

A Linear Relation

Based on the belief that starting salary is a function of grade-point average, we constructed a scattergram for the six data points available. This scattergram allowed us to determine that the two variables seem to be systematically associated. Furthermore, it appears that a linear function is a reasonable approximation of the relationship between these variables. Using Y_i to denote an individual's starting salary, and X_i to indicate the

Figure 2.2
Straight-line fit based on inspection of the data

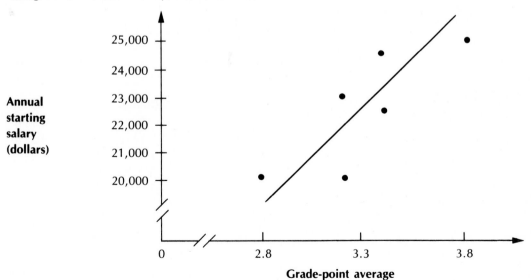

individual's corresponding grade-point average, we can now specify that on the average

$$\hat{Y}_i = a + bX_i$$

where a is the intercept (i.e., the estimated valued for Y_i when $X_i = 0$), b is the slope (i.e., the estimated change in Y_i for a one-unit increase in X_i), and $\hat{Y}_i$ (called "y-hat") is the predicted value for Y_i. The question is, how do we obtain values for a and b based on the six data points? One procedure is to draw the most reasonable straight line we can through the data in the scattergram. An example of such a drawing is provided in Figure 2.2.

If the graph were extended to include the origin, we could determine where the straight line intersects the vertical axis, to find the intercept. The slope coefficient is obtained by finding the change in annual starting salary that corresponds to a unit increase in grade-point average as indicated by the line. This is, however, quite a cumbersome procedure, and the results are influenced by any inaccuracy in drawing the straight line. Hence, a numerical procedure is likely to be more accurate and easier to use.

Least-Squares Method

Given data relevant to the problem on Y and X, the most common procedure for computing a, the intercept, and b, the slope coefficient, is the *least-squares* method. Once values for a and b are obtained we can compute a predicted value for Y, based on a given value for X; that is, let $\hat{Y}_i = a + bX_i$. The least-squares method involves choosing a and b so that the sum of the squared deviations

$$\sum_{i=1}^{n} (Y_i - \hat{Y}_i)^2$$

is minimized.[3] Graphically, this means that we choose the straight line that minimizes the sum of the squared vertical distances between the line and the dots in the scattergram. This is illustrated in Figure 2.3 for the observation that corresponds to the values $23,000 and 3.2. Now that the quantity to be minimized has been chosen (see Appendix 2.2 for the derivations), formulas for a and b can be obtained. These formulas are

$$b = \frac{\sum X_i Y_i - \dfrac{\sum X_i \sum Y_i}{n}}{\sum X_i^2 - \dfrac{(\sum X_i)^2}{n}}$$

$$a = \bar{Y} - b\bar{X}$$

Figure 2.3
Illustration of vertical distance for one observation

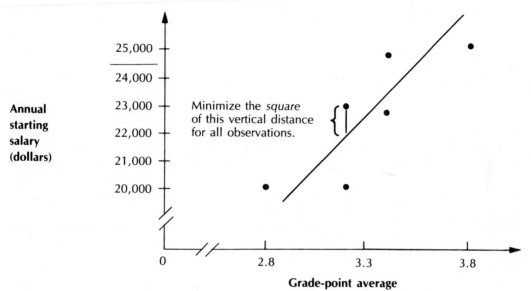

where

$$\bar{Y} = \frac{\Sigma Y_i}{n}$$

$$\bar{X} = \frac{\Sigma X_i}{n} \quad \text{and}$$

$$\Sigma = \sum_{i=1}^{n}$$

To perform the calculations it is convenient to organize the sample data as follows:

Y_i	X_i	Y_iX_i	X_i^2
20,000	2.8	56,000	7.84
24,500	3.4	83,300	11.56
23,000	3.2	73,600	10.24
25,000	3.8	95,000	14.44
20,000	3.2	64,000	10.24
22,500	3.4	76,500	11.56
135,000	19.8	448,400	65.88

$$b = \frac{448{,}400 - \dfrac{(19.8)(135{,}000)}{6}}{65.88 - \dfrac{(19.8)^2}{6}} = \frac{2{,}900}{0.54} = \$5{,}370 \text{ (rounded to nearest dollar)}$$

$$a = \frac{135{,}000}{6} - 5{,}370\left(\frac{19.8}{6}\right) = \$4{,}779$$

Thus, we have $\hat{Y}_i = 4{,}779 + 5{,}370X_i$ (on the average).

The literal interpretation of the intercept is that starting salary is predicted to be $4,779 when grade-point average is zero. From a substantive point of view this may not be very useful, because no student will graduate with such a grade-point average. However, the intercept is needed to completely specify the relationship.

Of primary interest is the slope coefficient, which indicates that for a one-unit increase in grade-point average, the predicted salary increases by $5,370. For example, for a grade-point average of 2.8 the predicted salary is $\hat{Y} = 4{,}779 + 5{,}370(2.8) = \$19{,}815$. A grade-point average of 3.8 gives $\hat{Y} = 4{,}779 + 5{,}370(3.8) = \$25{,}185$, or an additional $5,370.

These examples also illustrate how the equation can be used to predict starting salary based upon a grade-point average; however, these predictions are subject to error. Inaccuracies can arise partly because the intercept and slope coefficients are estimated from sample data and hence are subject to sampling error. Further, even if we had all possible data, the predictions would not be perfect because the relationship is not exact. To assess the statistical significance of our predictions, or of the intercept or slope coefficient, we need to make additional assumptions. This will be done in Chapter 3.

Problems

2.3 For the data in Problem 2.1, assume a linear relationship and compute the slope coefficient and the intercept.

2.4 For the data in Problem 2.2, assume a linear relationship and compute the slope coefficient and the intercept.

Goodness of Fit

With the sample data for Y and X, we can obtain a and b. With these estimates we can obtain *fitted* values for Y using the sample data. In the starting-salary example, we obtained

$$\hat{Y} = 4{,}779 + 5{,}370X$$

Using the sample data for X, we arrive at the following fitted values.

X_i	$\hat{Y}_i$
2.8	19,815
3.4	23,037
3.2	21,963
3.8	25,185
3.2	21,963
3.4	23,037

The closer these fitted values are to the actual values for Y in the sample, the better the model fits. The difference between an actual value (Y_i) in the sample and a fitted value ($\hat{Y}_i$) is called a *residual*, for which we use the notation $\hat{u}_i$ (called "u-hat"). The residuals and their squares are computed below.

Y_i	$\hat{Y}_i$	$\hat{u}_i$	$\hat{u}_i^2$
20,000	19,815	185	34,225
24,500	23,037	1,463	2,140,369
23,000	21,963	1,037	1,075,369
25,000	25,185	−185	34,225
20,000	21,963	−1,963	3,853,369
22,500	23,037	−537	288,369
		0	7,425,926

Note that the sum of the residuals equals zero. This is true by definition for the simple linear model using the least-squares method (see Appendix 2.3). Note also the calculations for the quantity $\Sigma \hat{u}_i^2$, and recall that a and b were chosen so as to minimize this quantity. Thus, given the sample data, all other values for a and b would result in a higher value for this sum.

We can use the sum of the squared residuals to determine the goodness of fit for the model. We want $\Sigma \hat{u}_i^2$ to be as close as possible to zero. However, it is difficult to gauge the goodness of fit by examining only this quantity, because it depends on the unit of measurement for the criterion variable. In other words, the absolute goodness of fit is influenced by how we measure the criterion variable, for example, in dollars or in thousands of dollars. One way to overcome this problem is to compare the sum of the squared residuals with the sum of the squared deviations from the sample mean.

Earlier in this chapter, we computed this latter quantity as $\Sigma(Y_i - \bar{Y})^2 = 23,000,000$. By comparing the sum of the squared residuals with this number, we get some idea of the potential improvement in estimating starting salary by using information on grade-point average. We illustrate these two quantities in Figures 2.4 and 2.5.

Figure 2.4
Total amount of variation

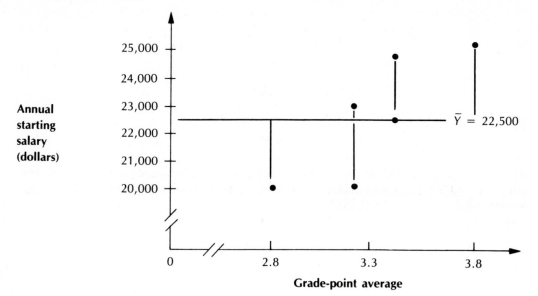

Figure 2.5
Unexplained amount of variation

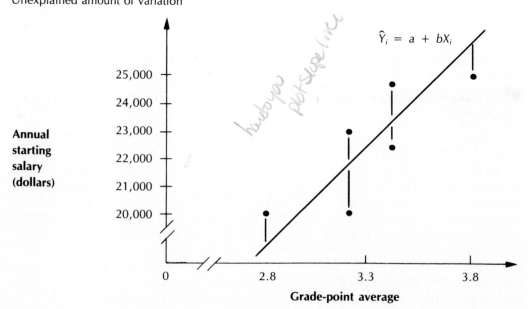

The total amount of variation in Y is shown in Figure 2.4, and the sum of the squared residuals (i.e., the unexplained amount of variation) is shown in Figure 2.5. From these two graphs we see that the distance between an observed value and the horizontal line (corresponding to $\bar{Y} = 22{,}500$) tends to be greater than the distance between the same observed value and the least-squares line.

Relative Goodness of Fit

To summarize the improvement achieved in estimating the criterion variable by using the least-squares line ($\hat{Y}$ line) instead of the horizontal line ($\bar{Y}$ line), it is customary to compute the *proportion* of the variation in the criterion variable accounted for by the model. To develop this measure, we separate the total amount of variation in the criterion variable into two parts: explained (accounted for) and unexplained (not accounted for).

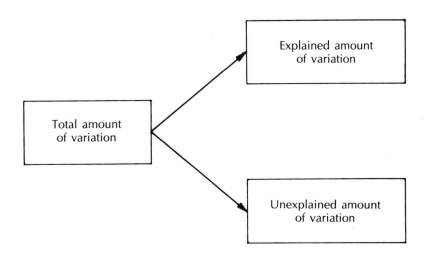

A relative measure of goodness of fit is obtained by taking the ratio of the explained variation over the total variation. Specifically,

$$R^2 = \frac{\text{Explained variation in } Y}{\text{Total variation in } Y}$$

Or since the unexplained variation equals the total variation minus the explained variation, we can also say

$$R^2 = 1 - \frac{\text{Unexplained variation in } Y}{\text{Total variation in } Y}$$

R^2 is often referred to as the *coefficient of determination.* Recall that $\Sigma(Y_i - \bar{Y})^2$ = 23,000,000 represents the total variation, and that $\Sigma(Y_i - \hat{Y}_i)^2$ = 7,425,926 is the unexplained variation. Thus,

$$R^2 = 1 - \frac{7,425,926}{23,000,000} = 1 - 0.323 = 0.677$$

In this case, R^2 indicates that the model accounts for approximately 68 percent of the variation in starting salary in the sample.

Formally, in equation form we have

$$\Sigma(Y_i - \bar{Y})^2 \quad = \quad \Sigma(\hat{Y}_i - \bar{Y})^2 \quad + \quad \Sigma(Y_i - \hat{Y}_i)^2$$

| Total amount of variation | Explained amount of variation | Unexplained amount of variation |

(See Appendix 2.4 for a proof of this equality.) R^2 is then expressed as

$$R^2 = \frac{\Sigma(\hat{Y}_i - \bar{Y})^2}{\Sigma(Y_i - \bar{Y})^2} = 1 - \frac{\Sigma(Y_i - \hat{Y}_i)^2}{\Sigma(Y_i - \bar{Y})^2}$$

For the simple linear model, the largest possible value is $R^2 = 1$, and the smallest is $R^2 = 0$. These indicate perfect fit and complete lack of fit respectively.

Problems

2.5 For the data in Problem 2.1, compute R^2 (the coefficient of determination) and interpret the result.

2.6 For the data in Problem 2.2, compute R^2 (the coefficient of determination) and interpret the result.

Correlation Coefficient

As long as we use a simple linear model (i.e., a model with only one predictor variable), R^2, the coefficient of determination, can also be written as r^2, the square of the correlation coefficient. This *correlation coefficient* can be computed directly as:

$$r = \frac{\Sigma X_i Y_i - \dfrac{\Sigma X_i \Sigma Y_i}{n}}{\sqrt{\Sigma X_i^2 - \dfrac{(\Sigma X_i)^2}{n}} \sqrt{\Sigma Y_i^2 - \dfrac{(\Sigma Y_i)^2}{n}}}$$

The correlation coefficient measures the degree of *linear* association between the two variables Y and X. A positive correlation coefficient indicates that as one variable increases in magnitude, the other variable also tends to go up in value. Conversely, a negative correlation coefficient indicates that as one variable goes up in magnitude, the other variable tends to go down in value. If the correlation coefficient is zero, there is no linear association between the two variables.

To compute the correlation coefficient, we need to calculate the measures in the formula shown above; however, to simplify the computations, we divide the salary figures by 1,000.[4] The calculations are

Y_i	X_i	Y_iX_i	Y_i^2	X_i^2
20	2.8	56	400	7.84
24.5	3.4	83.3	600.25	11.56
23	3.2	73.6	529	10.24
25	3.8	95	625	14.44
20	3.2	64	400	10.24
22.5	3.4	76.5	506.25	11.56
135	19.8	448.4	3060.50	65.88

Substituting these summary values into the formula for r we obtain:

$$r = \frac{448.4 - \dfrac{(19.8)(135)}{6}}{\sqrt{65.88 - \dfrac{(19.8)^2}{6}}\sqrt{3060.50 - \dfrac{(135)^2}{6}}} = \frac{2.9}{\sqrt{0.54}\sqrt{23}} = 0.82$$

To verify our previous calculations for R^2, we note that $r^2 = (0.82)^2 = 0.67$, which is identical (except for rounding errors) to $R^2 = 0.677$. Based on the computed value of the correlation coefficient of the sample data, we can assert that there is a substantial degree of positive linear association between the two variables.

The correlation coefficient is also related to the slope coefficient in simple linear regression. Specifically, it can be shown (see Appendix 2.6) that

$$r = b\frac{s_x}{s_y}$$

where b is the slope coefficient,

$$s_x = \sqrt{\frac{\Sigma(X_i - \bar{X})^2}{n - 1}}$$ is the standard deviation of X, and

$$s_y = \sqrt{\frac{\Sigma(Y_i - \bar{Y})^2}{n - 1}}$$ is the standard deviation of Y.

In other words, the correlation coefficient is equal to the slope coefficient, multiplied by the ratio of the standard deviation of the predictor variable (X_i) to the standard deviation of the criterion variable (Y_i). In a sense, we can think of the correlation coefficient as a *standardized* slope coefficient. Note the difference in interpretation of the slope and correlation coefficients for a linear regression of Y_i as a function of X_i:

Coefficient	Interpretation
Slope	Predicted unit change in Y for a one-unit increase in X
Correlation	Predicted *standard deviation* change in Y for a one *standard deviation* increase in X

Problems

2.7 For the data in Problem 2.1, compute r (the correlation coefficient) and interpret the result.

2.8 For the data in Problem 2.2, compute r (the correlation coefficient) and interpret the result.

Nonlinear Association

It is important to emphasize that the correlation coefficient is a measure of the *linear* association between two variables. Consider, for example, the following hypothetical data on annual household income and the amount of money paid for a new personal car:

Income (dollars)	Price (dollars)
10,000	5,000
50,000	6,000
50,000	9,000
10,000	10,000
30,000	13,000
30,000	14,000

We can verify that the correlation coefficient based on these data is $r = 0$ (i.e., there is no linear association between price and income); however, the scattergram for these data in Figure 2.6 reveals that the two variables are related.

The scattergram suggests that there may be a nonlinear relationship between the variables. Specifically, for the available hypothetical data, middle-income households ($30,000) spend more on a personal car than either low-income ($10,000) or high-income ($50,000) households. This hypothetical example illustrates the importance of using a scattergram to examine the association between variables to determine whether a linear function is appropriate.

Figure 2.6
Scattergram of data on car price and household income

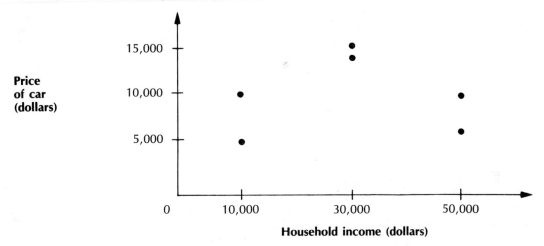

Use of Regression and Correlation Coefficients

Correlation coefficients are widely computed and used in published studies (e.g., in academic journals) and in studies conducted for firms or institutions. In many areas of application, a correlation coefficient receives more attention than a regression coefficient. Nevertheless, for managerial or policy decisions the regression (slope) coefficient is the preferred and most relevant measure. For example, economists find a great deal of use for regression coefficients in their work, while psychologists tend to emphasize correlation measures. Managerial or policy decisions are generally closer to the economist's perspective than they are to the psychologist's. We return to this topic in Chapter 8.

SUMMARY

In this chapter we introduced simple linear regression analysis. This technique is used to obtain a straight line which may describe the approximate linear relation between a criterion and a predictor variable, based on a sample of data. If we can justify the assumption of a linear relation, we can compute the slope coefficient and the intercept. The resulting equation can be used to predict the criterion variable based on a specified value for the predictor variable.

We emphasize that graphical methods such as scattergrams are important in determining whether a linear or nonlinear function is appropriate for the data. In Chapter 6 we discuss some ways in which nonlinear relations can be handled. We also recommend that the problem be studied from a substantive or theoretical point of view. In other words, we should always ask the question: does it make sense to assume a particular functional form to describe the relationship between the variables?

The least-squares method is the most frequently used procedure for obtaining a linear function. We have shown equations to obtain the slope coefficient and the intercept from sample data on two variables. Furthermore, we have derived and illustrated two measures, R^2, the coefficient of determination, and r, the correlation coefficient, which can be used respectively to summarize the fit of the linear equation and the degree of linear association between the variables. Finally, we emphasize that we do not need any additional assumptions if we only want to describe a relationship; however, if we want to use the equation to make forecasts or to apply our understanding of the relationship, we need statistical inference. We will take this up in the next chapter.

ADDITIONAL PROBLEMS

2.9 Product XYZ is manufactured by the Richmond Company once a month in lots which vary in size because of demand fluctuations. The data on lot size and work-hours of labor for ten production runs are shown below.

Production run	Work-hours	Lot size
1	73	30
2	50	20
3	128	60
4	170	80
5	87	40
6	108	50
7	135	60
8	69	30
9	148	70
10	132	60

a. Plot the data to show how the number of work-hours required depends on the lot size. Put work-hours on the vertical axis. Does it appear that there is linear relationship between the two variables?

b. Compute the linear relationship for the number of work-hours as a function of the lot size (i.e., compute the slope coefficient and the intercept) for the model:

$$\hat{Y}_i = a + bX_i$$

where Y_i = work-hours, and

X_i = lot size.

Interpret the results.

c. Compute the coefficient of determination and the correlation coefficient. Describe in plain English (to someone who does not know regression analysis) the interpretation of these results.

APPENDIX 2.1
DIRECT COMPUTATION OF CONDITIONAL MEANS

Given the following six data points,

Starting annual salary (dollars)	Grade-point average
20,000	2.8
24,500	3.4
23,000	3.2
25,000	3.8
20,000	3.2
22,500	3.4

we want to predict an annual salary based on grade-point average. To do so, we can summarize the six observations by computing the mean salary for each separate grade-point average value in the sample. For example, the smallest grade-point average value observed in the sample is 2.8, for which a starting salary of $20,000 is observed. The next smallest grade-point average is 3.2, for which the mean starting salary equals $21,500, and so forth. In this manner we obtain conditional mean starting salary values for all separate grade-point averages observed in the sample as follows:

Conditional mean starting salary (dollars)	Grade-point average
20,000	2.8
21,500	3.2
23,500	3.4
25,000	3.8

Starting salary can now be predicted by using this table of conditional mean values. Although we are not making any assumption about the form of the relationship, as is done using simple linear regression, this approach of developing conditional means has several limitations.

Limitations

1. We need several observations for each grade-point average before we can feel confident about the accuracy of the corresponding mean salary.
2. We need many observations over a wide variety of different grade-point averages before we can develop a table that covers a sufficient number of alternative possibilities to make the summary of conditional means interesting.
3. No matter how many observations for different grade-point averages we have, it is likely that we want to make predictions for values that are not included in a summary of conditional means. To do so, we will have to interpolate, which requires making assumptions about the form of the relationship.

Because of these and other limitations, regression analysis is usually considered to be superior to the direct computation of conditional means, *as long as* we verify the appropriateness of the functional form and recognize that the relationship is an approximation which may not extend far beyond the range of values for the predictor variable observed in the sample. However, if we use our knowledge of the problem to specify the most appropriate functional form for each relevant predictor variable, the estimated equation may be applicable in general.

APPENDIX 2.2
DERIVATIVES TO SOLVE FOR A AND B

Minimize

$$\Sigma(Y_i - \hat{Y}_i)^2$$

or minimize

$$\Sigma(Y_i - a - bX_i)^2, \quad \text{since } \hat{Y}_i = a + bX_i.$$

Take partial derivatives of this function with respect to a and b, and set the resulting equations equal to zero:

$$\frac{\partial \Sigma(Y_i - a - bX_i)^2}{\partial a} = -2\Sigma(Y_i - a - bX_i) = 0$$

$$\frac{\partial \Sigma(Y_i - a - bX_i)^2}{\partial b} = -2\Sigma X_i(Y_i - a - bX_i) = 0$$

Simplifying the two equations with two unknowns

$$\Sigma Y_i - \Sigma a - b\Sigma X_i = 0$$

$$\Sigma X_i Y_i - a\Sigma X_i - b\Sigma X_i^2 = 0$$

or

$$\Sigma Y_i = na + b\Sigma X_i$$

$$\Sigma X_i Y_i = a\Sigma X_i + b\Sigma X_i^2$$

And solving for a and b we obtain

$$b = \frac{\Sigma X_i Y_i - \dfrac{\Sigma X_i \Sigma Y_i}{n}}{\Sigma X_i^2 - \dfrac{(\Sigma X_i)^2}{n}}$$

$$a = \bar{Y} - b\bar{X}$$

where $\qquad \bar{Y} = \dfrac{\Sigma Y_i}{n} \qquad$ and

$$\bar{X} = \dfrac{\Sigma X_i}{n}$$

APPENDIX 2.3
PROOF THAT RESIDUAL SUM IS ZERO

To show

$$\Sigma \hat{u}_i = 0$$

recall that

$$\begin{aligned}
\hat{u}_i &= Y_i - \hat{Y}_i \\
&= Y_i - (a + bX_i) \\
&= Y_i - (\bar{Y} - b\bar{X} + bX_i)
\end{aligned}$$

then

$$\begin{aligned}
\Sigma \hat{u}_i &= \Sigma Y_i - \Sigma \bar{Y} + b\Sigma \bar{X} - b\Sigma X_i \\
&= \Sigma Y_i - n\bar{Y} + bn\bar{X} - b\Sigma X_i \\
&= \Sigma Y_i - n\frac{\Sigma Y_i}{n} + bn\frac{\Sigma X_i}{n} - b\Sigma X_i \\
&= \Sigma Y_i - \Sigma Y_i + b\Sigma X_i - b\Sigma X_i \\
&= 0
\end{aligned}$$

APPENDIX 2.4
PROOF THAT VARIATION IN *Y* EQUALS EXPLAINED
PLUS UNEXPLAINED VARIATION

To show

$$\Sigma(Y_i - \bar{Y})^2 = \Sigma(\hat{Y}_i - \bar{Y})^2 + \Sigma(Y_i - \hat{Y}_i)^2$$

Proof:

$$\begin{aligned}
\Sigma(Y_i - \bar{Y})^2 &= \Sigma(Y_i - \hat{Y}_i + \hat{Y}_i - \bar{Y})^2 \\
&= \Sigma(Y_i - \hat{Y}_i)^2 + 2\Sigma(Y_i - \hat{Y}_i)(\hat{Y}_i - \bar{Y}) + \Sigma(\hat{Y}_i - \bar{Y})^2
\end{aligned}$$

Note that this expression differs from the starting one only in the second term. If the second term is zero, then we have completed the proof. Thus, we have to show

$$\Sigma(Y_i - \hat{Y}_i)(\hat{Y}_i - \bar{Y}) = 0$$

Substituting

$$Y_i = a + bX_i + \hat{u}_i$$
$$\hat{Y}_i = a + bX_i$$
$$\bar{Y} = a + b\bar{X}$$

we have

$$\Sigma(a + bX_i + \hat{u}_i - a - bX_i)(a + bX_i - a - b\bar{X})$$
$$= \Sigma(\hat{u}_i)b(X_i - \bar{X})$$
$$= b\Sigma\hat{u}_iX_i - b\bar{X}\Sigma\hat{u}_i$$
$$= b\Sigma\hat{u}_iX_i \qquad \text{since } \Sigma\hat{u}_i = 0$$
$$\text{(see Appendix 2.3)}$$
$$= 0 \qquad \text{since } \Sigma\hat{u}_iX_i = 0$$
$$\text{(see Appendix 2.5)}$$

APPENDIX 2.5
PROOF OF ZERO COVARIANCE BETWEEN RESIDUALS AND PREDICTOR VARIABLE

To show

$$\Sigma\hat{u}_iX_i = 0$$

Proof:

$$\Sigma\hat{u}_iX_i = \Sigma(Y_i - a - bX_i)X_i \qquad \text{since}$$
$$\hat{u}_i = Y_i - a - bX_i$$
$$= \Sigma(Y_i - (\bar{Y} - b\bar{X}) - bX_i)X_i \qquad \text{since}$$
$$a = \bar{Y} - b\bar{X}$$
$$= \Sigma(Y_i - \bar{Y})X_i - b\Sigma(X_i - \bar{X})X_i$$

$$= \left(\Sigma X_i Y_i - \frac{\Sigma Y_i \Sigma X_i}{n} \right) - b \left(\Sigma X_i^2 - \frac{(\Sigma X_i)^2}{n} \right)$$

since

$$\bar{Y} = \frac{\Sigma Y_i}{n} \quad \text{and}$$

$$\bar{X} = \frac{\Sigma X_i}{n}$$

$$= 0$$

since

$$b = \frac{\Sigma X_i Y_i - \dfrac{\Sigma X_i \Sigma Y_i}{n}}{\Sigma X_i^2 - \dfrac{(\Sigma X_i)^2}{n}}$$

APPENDIX 2.6
RELATION BETWEEN SLOPE
AND CORRELATION COEFFICIENTS

For a simple linear regression, the correlation coefficient equals the slope coefficient multiplied by the ratio of the standard deviations for X over Y.

Proof:

$$b = \frac{\Sigma Y_i Y_i - \dfrac{\Sigma X_i \Sigma Y_i}{n}}{\Sigma X_i^2 - \dfrac{(\Sigma X_i)^2}{n}}$$

$$b \, \frac{s_x}{s_y} = \frac{\Sigma X_i Y_i - \dfrac{\Sigma X_i \Sigma Y_i}{n}}{\Sigma X_i^2 - \dfrac{(\Sigma X_i)^2}{n}} \cdot \frac{\sqrt{\left(\Sigma X_i^2 - \dfrac{(\Sigma X_i)^2}{n} \right) \Big/ (n-1)}}{\sqrt{\left(\Sigma Y_i^2 - \dfrac{(\Sigma Y_i)^2}{n} \right) \Big/ (n-1)}}$$

$$= \frac{\Sigma X_i Y_i - \dfrac{\Sigma X_i \Sigma Y_i}{n}}{\sqrt{\Sigma X_i^2 - \dfrac{(\Sigma X_i)^2}{n}} \sqrt{\Sigma Y_i^2 - \dfrac{(\Sigma Y_i)^2}{n}}}$$

$$= r$$

ENDNOTES

1. Of course, if you feel you are better than the average fellow student, you may want to use a number higher than the average starting salary. This is similar to the notion that the salary depends on certain factors. Regression analysis is especially useful to accommodate such factors.

2. See Appendix 2.1 for an alternative approach to the problem of predicting salary conditional upon grade-point average.

3. Remember that computing a squared deviation makes sense only if Y, the criterion variable, is measured on at least an interval scale.

4. The correlation coefficient is not affected by such a transformation of the data on the criterion or predictor variables.

3

Statistical Inference in Simple Regression Analysis

Chapter 2 showed how we obtain least-squares estimates for the slope coefficient and intercept in a linear model with one predictor variable. The resulting equation after obtaining these estimates can be used to predict a value for the criterion variable given a value for the predictor variable, and to understand how variables relate to each other. For either purpose the results' statistical reliability (i.e., statistical significance) should also be established. To do so, the uncertainty associated with the estimates must be quantified. Before we discuss statistical inference in regression analysis, we review statistical inference about a population mean.

Statistical Inference about a Mean

To facilitate the discussion about statistical inference for regression analysis, we first review the procedures and assumptions for statistical inference about the mean of a population. Let Y_i denote the values for a random variable, and $i = 1, 2, \ldots, N$ are the number of observations in the population. To estimate the mean, μ, of this population, we may use a simple random sample $i = 1, 2, \ldots, n$, from the population. (Every possible sample of size n has an equal chance of being drawn from the population.) The use of simple random sampling implies that

1. $E(Y_i) = \mu$ for all i

2. $\text{Var}(Y_i) = \sigma^2$ for all i

3. $\text{Cov}(Y_i, Y_j) = 0$ for $i \neq j$

In words, (1) the expected value for each separate draw from the population equals the mean value μ; (2) the variance of the random variable equals σ^2; and (3) the covariance between the results from two separate drawings is zero. See any basic statistics book for a discussion about variance and covariance.

In more concrete terms, consider a population that contains only observations with values equal to 1, 2, 3, 4, 5, and 6. Each possible value occurs with the same frequency. That is, the number of ones equals the number of two's, etc. The population resembles a die, with the number of observations in the population being determined by the number of tosses. The mean of this population is 3.5, and the variance can be found to be 2.92 (see Appendix 3.1). Also, the covariance is zero, which can be interpreted as indicating that the result of one drawing (or one toss of the die) has no bearing on the possible outcome of the next, or any other, drawing (or the next toss of the die).

Suppose we take one simple random sample of size n. And let

$$\bar{Y} = \frac{\sum_{i=1}^{n} Y_i}{n}$$

be the formula for the sample mean. Then it can be shown that[1]

1. $E(\bar{Y}) = \mu$

2. $\text{Var}(\bar{Y}) = \sigma_{\bar{Y}}^2 = \frac{\sigma^2}{n}$

These two results suggest that (1) the sample mean, using simple random sampling is *unbiased* (the expected value equals the population value), and (2) the variance of the sample mean becomes smaller as the sample size increases (based on the denominator in the formula for the variance). In addition, it can be shown that the sample mean is an *efficient* or minimum-variance estimator (i.e., no other estimator of the population mean has a smaller variance).

Finally, if $\bar{X}$, as a random variable, is normally distributed,[2] we can use the standard normal distribution to test hypotheses or to construct confidence intervals for the population mean. Specifically, Z, the standard normal value, can be computed as follows:

$$Z = \frac{\bar{Y} - \mu}{\sigma_{\bar{Y}}}$$

where $\bar{Y}$ is the calculated sample mean,

μ is the hypothesized population mean, and

$\sigma_{\bar{Y}}$ is the standard error of the mean (or the standard deviation of the theoretical distribution of all possible sample mean values for a given sample size).

Given that

$$\sigma_{\bar{Y}} = \frac{\sigma}{\sqrt{n}}$$

with σ being usually unknown, we tend to use the *t*-distribution instead of the standard normal distribution. The *t*-value is computed using the formula

$$t = \frac{\bar{Y} - \mu}{s_{\bar{Y}}}$$

where

$$s_{\bar{Y}} = \frac{s}{\sqrt{n}} \quad \text{and}$$

$$s = \sqrt{\frac{\Sigma(Y_i - \bar{Y})^2}{n - 1}}$$

Given the denominator in the formula for s, the appropriate *t*-distribution is the one with $(n - 1)$ degrees of freedom.

Statistical Inference in Regression Analysis

For the investigation of relationships, using statistical inference, we need a justification similar to the one for inferences about a population mean. However, one important difference is that we do *not* use random sampling. Instead, we tend to use whatever data we have available for estimating and testing relationships between variables (recall the example of starting salaries). Thus, instead of being able to say that "random sampling implies ..." we state certain *assumptions* that are equivalent to the implications resulting from random sampling. Of course, these assumptions must be justified, and much of our discussion in subsequent chapters (especially Chapter 7) centers on this issue. In this chapter, we state the assumptions, and indicate their role in statistical inference about the parameters in a regression equation as well as in inferences about predicted values.

  To make the analogy with the implications from random sampling for the random variable Y_i, we first express in a different form the results of random drawings from the population as

$$Y_i = \mu + u_i$$

where u_i is an error term which captures the (random) difference between a particular value for Y_i and the population mean. The use of simple random sampling has the following implications for this error term $(u_i = Y_i - \mu)$:

1. $E(u_i) = 0$ for all i

2. $Var(u_i) = \sigma^2$ for all i

3. $Cov(u_i, u_j) = 0$ for $i \neq j$

In words, (1) the expected value of the error term is zero (this follows directly from the implication that the expected value of the random variable Y_i equals the population mean); (2) the variance of the error term is σ^2, which equals the variance of Y_i because μ is a constant, and hence

$$\text{var}(u_i) = \text{var}(Y_i - \mu) = \text{var}(Y_i) = \sigma^2$$

and (3) the covariance of two different error terms is the same as the covariance of two different drawings from the population (because μ is a constant, and hence $\text{cov}(u_i, u_j) = \text{cov}(Y_i, Y_j)$).

Now recall the equation for a linear relation between Y, the criterion variable, and X, the predictor variable:

$$\hat{Y}_i = a + bX_i$$

This equation includes an intercept, a, and a slope coefficient, b, which are to be obtained from a set of data. This equation can be used as an estimate of the "population" equation

$$Y_i = \alpha + \beta X_i + u_i$$

where α is the unknown population intercept value,

           β is the unknown population slope value or slope parameter, and

           u_i is the error term associated with the ith observation.

Reasons the Relationship is Not Exact

We have adopted the simple linear model

$$Y_i = \alpha + \beta X_i + u_i$$

In this model, α and β are unknown parameters which can be estimated from sample data. The error term in this model, u_i, is also unobservable, but it can be estimated from sample data. This unobservable error is the reason for the inexact or *stochastic* relationship. More precisely, inexactness exists for several reasons.

First, most problems in management and economics (and the social sciences in general) involve complex relationships between variables. It is quite unrealistic to expect a perfect (i.e., completely deterministic) linear relation between two variables. Usually there are many variables which relate to the criterion variable. To model relationships we use only a small number of predictor variables (so far only one) which we believe can account for much of the variation in the criterion variable. Therefore, one reason the error term exists is that it accounts for any *omitted* variables. However, any single omitted variable by itself should have only a slight effect on the criterion variable. The problem of omitted variables is discussed in some detail in Chapter 4.

Second, even if all relevant variables were included in the model, the functional form of the relationship between the criterion and predictor variables may differ from the one chosen. Obviously the model will not be very useful if the correct form of the relationship differs dramatically from the one chosen; however, if the chosen form is a reasonable approximation of the actual relationship, the model can be useful for the purposes of prediction and understanding. Nevertheless, this approximation is another reason for the error term. We return to this problem of choosing the functional form in Chapter 6.

Third, the measurement of the criterion and predictor variables may be imperfect. For example, assume that Y, the criterion variable, is measured inaccurately because the measuring instrument is imprecise. If such measurement errors in Y are unrelated to the predictor variable(s), then they cannot be explained. Hence, the greater such measurement errors, the greater the error term in the model.

Fourth, data are often available only at some aggregate level. If the relationship differs between individuals or if it changes over time, we should attempt to accommodate such differences in the modeling effort. However, if such differences or changes are minor or nonsystematic, we may assume a "fixed" relationship. In that case, the error term in the model may reflect differences between individuals, or changes in the relationship over time.

Fifth, given that the data we use are based on human behavior, the error term in the model may account for a "random" component in behavior. Although this is a controversial point in some circles, there is no question that part of human behavior is unpredictable or is predictable only with very complex models; thus, from a practical point of view, it is reasonable to argue that the error term in the model exists partly because we cannot predict human behavior with complete accuracy.

Assumptions about the Error Term

Based on a set of available data, we can compute a and b as estimates of α and β. Quantifying the statistical uncertainty inherent in these estimates, and consequently the uncertainty about any predictions based on the estimated equation, requires us to make *assumptions* about the error term. Following is a list of specific assumptions (note that random sampling from a population provides similar results):

ASSUMPTION 1: $E(u_i) = 0$ **for all** i**, given a value for** X**.**

The first assumption is that the error term has an expected value of zero for each observation. The specific value for this error may vary from one observation to another, but on the average its value is assumed to be zero. This implies that the expected value of Y_i*, given some value for* X_i*, equals* $\alpha + \beta X_i$*:*

$$E(Y_i|X_i) = \alpha + \beta X_i \qquad \text{for all } i.$$

(handwritten margin note: ? how can it be o really *)*

ASSUMPTION 2: $\text{Var}(u_i) = \sigma^2$ **for all** i**, given a value for** X**.**

The second assumption is that the unobservable variance of the error term for each observation equals an unknown constant σ^2*. This variance describes the amount of variability in the possible error values associated with an observation. It is assumed to always equal* σ^2*. Another way to look at this, is to say that for any given value of* X*, the variance of* Y *is equal to* σ^2*, or* $\text{Var}(Y_i|X_i) = \text{Var}(u_i) = \sigma^2$*. This assumption is often referred to as homoscedasticity, while a violation of the assumption is referred to as heteroscedasticity.*

The first two assumptions are summarized in Figure 3.1. The solid line indicates the true relation $(\alpha + \beta X)$, on the average, between Y and X. Parallel to and equally distant from the solid line are two dotted lines. According to the assumptions, the number of observations between the solid line and the upper dotted line is expected to be equal to the number of observations between the solid line and the lower dotted line.

Figure 3.1
Summary of error-term assumptions (1) and (2)

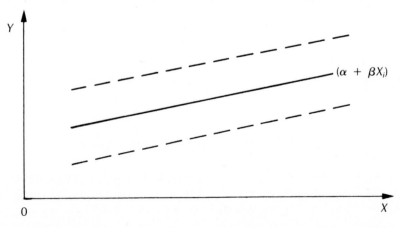

ASSUMPTION 3: $Cov(u_i, u_j) = 0$ **for** $i \neq j$, **given any two values for** X.

The third assumption is that the possible values for the error terms for any two different observations have zero covariance. Remember that u_1, u_2, ..., u_n are random variables. The processes that generate values for these random variables are assumed to be independent of each other; thus, whatever value is obtained for the error term for one observation is assumed not to be related to the value of the error term for any other observation in the sample.[3]

ASSUMPTION 4: u_i **is distributed according to the normal distribution for all** i, **given a value for** X_i.

The fourth assumption is that there is a bell-shaped distribution for all possible error-term values. This implies, among other things, that the absolute value of the error term is more likely to be small than large.

To summarize the assumptions, the error term, u_i, is independently and normally distributed with a mean of zero and a variance of σ^2 for each observation given a value for X. Figure 3.2 graphically summarizes all four assumptions. The straight line in this figure corresponds to the average relation between Y and X. The interpretation of the two curves, at $X = X_1$ and $X = X_2$, requires three dimensions. For any point along the X-axis, the errors in the relation are randomly distributed according to the normal distribution with mean zero and variance σ^2. Also, the outcomes of the random drawings for the error terms for any two values of X are independent.

Understanding as the Objective

If we use regression analysis to understand how a predictor variable relates to a criterion variable, we often assume that X causes Y. That is, we assume that the value for Y is

Figure 3.2
Summary of error-term assumptions (1) through (4)

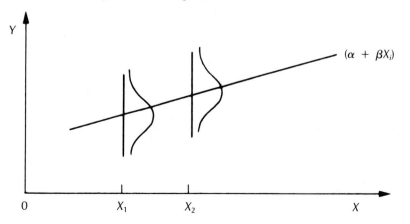

determined, at least in part, by the value for X. For example, if we estimate the relation between starting salary and grade-point average, we may do so because we implicitly believe that a higher grade-point average causes an increase in starting salary.

Symbolically, the requirement of causality can be represented as

$$Y \leftarrow X$$

This requirement, strictly speaking, is important *only* when the purpose of the study is to understand the relationship. It implies that if we control X, we can determine, in part, the value for Y. If in fact Y is determined in a different manner from what is assumed, a different model must be specified before a useful understanding of the relationship can be gained through regression analysis. For example, imagine that we have measured three variables, X, Y, and Z. Now, suppose that Z influences X and that Z also influences Y, as shown here.

Even if X and Y are not directly related to each other, we may observe a substantial correlation (association) between them; however, in this case the observed association does not result from a *causal* relationship between X and Y. Only prior knowledge and understanding of the actual problem will help us recognize that Z exists; regression analysis will not be helpful in distinguishing between causation and association.

If the purpose of the regression analysis is to understand relationships so that outcomes can be controlled (for example, by manipulating the predictor variable), we are implicitly assuming a causal relationship. Unfortunately, regression analysis cannot help us determine whether this assumption is reasonable. However, if we want strong evidence of a causal relationship, we can manipulate the predictor variable in a systematic and controlled manner to provide such evidence. This is often done through experimentation.

Difference between Errors and Residuals

We use the error term to refer to the unobservable, random component, u_i, in the model. This term is unobservable because it equals

$$Y_i - \alpha - \beta X_i$$

and α and β are unobservable. Given an estimated equation,

$$\hat{Y}_i = a + bX_i$$

we can observe and compute the difference between a value for the criterion variable in the sample, Y_i, and the corresponding *fitted* value, $\hat{Y}_i$. That is, let the residual be

$$\hat{u}_i = Y_i - \hat{Y}_i$$

Then $\hat{u}_i$, a residual, may be used as an estimate of the unobservable error term. In practice, we use the residuals to investigate whether there is evidence to doubt the validity of the assumptions made about the error term. There are, however, certain limitations associated with the use of residuals instead of the unobservable error terms. For example, before we can compute the residuals, we have to estimate two parameters, α and β. This estimation results in the loss of two degrees of freedom. With a sample size of n, the remaining degrees of freedom associated with the residuals is now $(n - 2)$.

Hypothesis Testing about β

Test statistics to infer the slope parameter, the intercept parameter, and other unknown population quantities can be derived using the assumptions described earlier. To keep the discussion simple, we present the formulas in stages; however, we do not present a full justification for the formulas to avoid mathematical details.

In statistics, it is desirable to have an estimator that is unbiased and efficient. Now, if Assumption (1) about the error term is true, then b is an unbiased estimator of β:

$$E(b) = \beta$$

Furthermore, if Assumptions (2) and (3) hold true, b is an efficient[4] (i.e., a minimum-variance) estimator of β. Therefore, together the first three assumptions imply that b is the best linear unbiased estimator ($b.l.u.e.$).

With these three assumptions we can derive a formula for the standard error of b:

$$\sigma_b = \frac{\sigma}{\sqrt{\Sigma(X_i - \bar{X})^2}}$$

where σ is the standard deviation of the error term, and
$\Sigma(X_i - \bar{X})^2$ is the sum of squared differences between observed values for X and the mean value for X in the sample.

And with Assumption (4) it follows that

$$\frac{b - \beta}{\sigma_b}$$

is distributed according to the standard normal distribution (i.e., the Z distribution in which $\mu = 0$ and $\sigma^2 = 1$). Thus, a computed Z-value can be obtained based on a computed value for b from sample data, a hypothesized value for β (under the null hypothesis), and a value for σ_b, the standard error of b. The formula for σ_b, however, requires knowing σ, which is the *unobservable* standard deviation of the error term. In practice, we compute s, the standard deviation of the residuals from the regression, and use it to obtain the *estimated* standard error of b as follows:

$$s_b = \frac{s}{\sqrt{\Sigma(X_i - \bar{X})^2}}$$

To accommodate the additional uncertainty from using s_b as an estimate of σ_b, we have to use the t-distribution. Specifically, it can be shown that (given the four assumptions)

$$\frac{b - \beta}{s_b}$$

follows the t-distribution with $n - 2$ degrees of freedom.

Standard Deviation of Residuals

Recall that a residual is the difference between an actual value (Y_i) in the sample and the fitted value ($\hat{Y}_i$) obtained from the equation. In our example we have the following data:

Y_i	$\hat{Y}_i$	$\hat{u}_i$	$\hat{u}_i^2$
20,000	19,815	185	34,225
24,500	23,037	1,463	2,140,369
23,000	21,963	1,037	1,075,369
25,000	25,185	−185	34,225
20,000	21,963	−1,963	3,853,369
22,500	23,037	−537	288,369
		0	7,425,926

The standard deviation of the residuals is defined as

$$s = \sqrt{\frac{\Sigma(\hat{u}_i - \hat{\bar{u}})^2}{n - 2}}$$

where $\hat{\bar{u}}$ is the average residual value.[5] However, since the sum of the residual values equals zero (see Appendix 2.3), the average residual value in the sample equals zero. Hence, the standard deviation of the residuals can be written as

$$s = \sqrt{\frac{\Sigma\hat{u}_i^2}{n - 2}} = \sqrt{\frac{\Sigma(Y_i - \hat{Y}_i)^2}{n - 2}}$$

In our example we have

$$s = \sqrt{\frac{7,425,926}{4}} \cong 1,363$$

This result can be interpreted as follows. If the assumptions about the (unobservable) error terms in the model are satisfied, then the (observable) residuals will follow approximately a normal distribution with a mean of zero (i.e., the average residual value is zero) and an estimated standard deviation of $1,363. Hence, approximately 68 percent of the residuals in the sample are expected to be between $-$1,363 and $+$1,363, and 95 percent between $-$2,726 and $+$2,726.

Problems

3.1 Compute the standard deviation of the residuals from the equation in Problem 2.3 and the data in Problem 2.1.

3.2 Compute the standard deviation of the residuals from the equation in Problem 2.4 and the data in Problem 2.2.

Testing a Hypothesis about the Slope

We may suggest the following eight-step procedure to test a hypothesis about the slope parameter. We assume that a two-tailed test is appropriate, and we use the sample data on starting salary and grade-point average to illustrate the procedure. It is possible to justify a one-tailed test by claiming a priori (prior to data analysis) that the only possible relation is, say, a positive one.

1. Establish the null and alternative hypotheses.
   a. The null hypothesis is that there is no linear relationship between Y and X, that is, $H_o : \beta = 0$
   b. The alternative hypothesis is that there is a linear relationship between Y and X, i.e. $H_A : \beta \neq 0$.
2. Determine the tolerance for a type I error probability. Allow the probability to be 5 percent; that is, if we reject the null hypothesis, the probability of being wrong in doing so is no more than 5 percent.
3. Identify an appropriate test-statistic. Using $H_o : \beta = 0$ the test statistic is

$$t = \frac{b - \beta}{s_b} = \frac{b}{s_b}$$

4. State the assumptions under which this test statistic is valid. The four assumptions about the error term must be true, to apply the test statistic.
5. Determine whether these assumptions are satisfied. Tests and graphical procedures are available to see if there is any evidence against one or more assumptions. These procedures are discussed in subsequent chapters. We proceed in this chapter as if there is no evidence against these assumptions.
6. Find the values for the test statistic that allow a rejection of the null hypothesis. Using the t-distribution with $n - 2$ degrees of freedom (df), we find the critical value to be 2.78 (4 df, 5 percent error, two tails). That is, we reject the null hypothesis if the computed t-value exceeds 2.78 or is less than -2.78. (See the t-distribution in Figure 3.3.)
7. Compute the t-value based on sample data. We need values for b and s_b. The slope coefficient, b, was found to be 5,370. The formula for s_b is

$$s_b = \frac{s}{\sqrt{\Sigma(X_i - \bar{X})^2}}$$

X_i	$(X_i - \bar{X})$	$(X_i - \bar{X})^2$
2.8	−0.5	0.25
3.4	0.1	0.01
3.2	−0.1	0.01
3.8	0.5	0.25
3.2	−0.1	0.01
3.4	0.1	0.01
		0.54

Figure 3.3
t-distribution

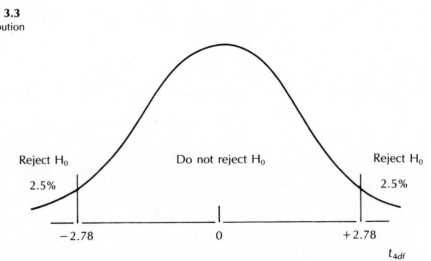

Then $\qquad\qquad s_b = \dfrac{1{,}363}{\sqrt{0.54}} \cong 1{,}854$ and

$$t = \frac{b - \beta}{s_b} = \frac{5{,}370 - 0}{1{,}854} \cong 2.90$$

8. Interpret the result from a statistical as well as a substantive viewpoint.
   a. Since the computed t-value of 2.90 exceeds the tabulated (critical) t-value of 2.78, we can reject the null hypothesis at the 5 percent significance level.
   b. We therefore conclude that there is a statistically significant linear relationship between starting salary and grade-point average.

Confidence Interval around *b*

Instead of testing the null hypothesis for the slope parameter, we may use a confidence interval to quantify the uncertainty associated with the slope coefficient. If the four assumptions about the error term are valid, the appropriate value from the t-table can be used to obtain a confidence interval for the slope parameter:

$$b \pm (t)(s_b)$$

For example, the tabulated value for a 95-percent confidence interval and 4 degrees of freedom is 2.78. For the sample data on starting salary and grade-point average we have

$$5{,}370 \pm (2.78)(1{,}854) \quad \text{or}$$

$$5{,}370 \pm 5{,}154$$

Thus the 95-percent confidence interval is 216 to 10,524. This confidence interval indicates the uncertainty about the magnitude of the slope parameter in the relationship between grade-point average and starting salary.

While we found, using the hypothesis-testing procedure, that the slope coefficient was statistically significantly different from zero, the confidence interval suggests nevertheless that there is a great deal of uncertainty about the exact magnitude of the slope parameter.

Problems

3.3 Test the null hypothesis that $\beta = 0$ for the equation in Problem 2.3. Use a significance level of 10 percent.

3.4 Test the null hypothesis that $\beta = 0$ for the equation in Problem 2.4. Use a significance level of 5 percent.

3.5 Construct a 98-percent confidence interval based on the slope coefficient computed in Problem 2.3.

3.6 Construct a 90-percent confidence interval based on the slope coefficient computed in Problem 2.4.

Hypothesis Testing about α

Hypothesis testing about α is quite similar to hypothesis testing about β. First, using Assumption (1) for the error term, it can be shown that a is an unbiased estimator of α, that is, $E(a) = \alpha$. Furthermore, if Assumptions (2) and (3) hold, a is an efficient estimator of α. Together the three assumptions imply that a is the best linear unbiased estimator ($b.l.u.e.$) of α.

Using these three assumptions, it can be shown that the formula for the standard error of a is

$$\sigma_a = \sigma \sqrt{\frac{1}{n} + \frac{\bar{X}^2}{\Sigma(X_i - \bar{X})^2}}$$

And with the fourth assumption we state that

$$\frac{a - \alpha}{\sigma_a}$$

is distributed according to the standard normal distribution. However, since σ_a is unobservable, we instead use the estimated standard deviation of a:

$$s_a = s \sqrt{\frac{1}{n} + \frac{\bar{X}^2}{\Sigma(X_i - \bar{X})^2}}$$

And it can be shown that the substituted formula

$$\frac{a - \alpha}{s_a}$$

follows the t-distribution with $(n - 2)$ degrees of freedom.

In our example

$$s_a = 1{,}363 \sqrt{\frac{1}{6} + \frac{(3.3)^2}{0.54}} \cong 6{,}146$$

To test the null hypothesis, $\alpha = 0$, against the alternative hypothesis, $\alpha \neq 0$, we compute

$$t = \frac{4{,}779 - 0}{6{,}146} = 0.78$$

For a 5-percent probability of a type I error, the tabulated t-value with 4-degrees of freedom is 2.78. Clearly, we cannot reject the null hypothesis that the intercept is zero.

Confidence Interval around *a*

A confidence interval may be constructed to quantify the uncertainty associated with the intercept. With the four assumptions about the error term the formula is

$$a \pm (t)(s_a)$$

Again using 95-percent confidence, we obtain

$$4{,}779 \pm (2.78)(6.146)$$

or

$$4{,}779 \pm 17{,}086$$

The 95-percent confidence interval is $-12{,}307$ to $21{,}865$, that is, the lower bound is $-\$12{,}307$ and the upper bound is $\$21{,}865$.

Problems

3.7 Test the null hypothesis that $\alpha = 0$ for the equation in Problem 2.3. Use a significance level of 10 percent.

3.8 Test the null hypothesis that $\alpha = 0$ for the equation in Problem 2.4. Use a significance level of 5 percent.

3.9 Construct a 98-percent confidence interval based on the intercept computed in Problem 2.3.

3.10 Construct a 90-percent confidence interval based on the intercept computed in Problem 2.4.

Hypothesis Testing of Correlation

If we only want to establish that a statistically significant degree of linear association exists between two variables, we can test the null hypothesis about the correlation. Keep in mind, of course, that linear association does not necessarily imply causality. And a statistically significant degree of linear association does not offer proof of causality. Note also that the correlation coefficient does not change if we reverse the criterion and predictor variables. For the null hypothesis that ρ, the population correlation, is zero, the test-statistic is

$$t = \frac{r\sqrt{n-2}}{\sqrt{1-r^2}}$$

In our example for the linear association between starting salary and grade-point average we obtained $r = 0.82$ (see Chapter 2). The computed value for the test statistic is

$$t = \frac{0.82\sqrt{4}}{\sqrt{1 - (0.82)^2}} = 2.87$$

For a two-tailed null hypothesis and a 5-percent type I error probability we can reject the null hypothesis since 2.87 exceeds 2.78, the tabulated t-value for 4 degrees of freedom. Therefore there is a statistically significant degree of linear association between the two variables.[6]

Problems

3.11 Test the null hypothesis that ρ, the population correlation coefficient, is zero, using the computed correlation coefficient in Problem 2.7.

3.12 Test the null hypothesis that ρ, the population correlation coefficient, is zero, using the computed correlation coefficient in Problem 2.8.

Prediction as the Objective

If our main purpose in conducting a regression analysis is to predict the criterion variable, we may also prefer to have a causal relation between Y and X (i.e., $Y \leftarrow X$), but this is not essential. To clarify this, consider two alternative prediction scenarios.

Prediction Scenario 1

If we manipulate X, and we want to predict Y values for alternative X values, we must assume causality.

Prediction Scenario 2

If we do not have any control over X, and we want to predict the value of Y given some value for X, causality is not needed as long as any other conditions that influence the observed association between Y and X are held constant. In practice, however, there is no way to guarantee that these other conditions are held constant.

Goodness of Fit

To summarize the equation's performance we may use either an absolute or a relative measure of goodness of fit. An absolute measure of fit is the standard deviation of residuals, s, which estimates the standard deviation of the error term, σ. The computed value, in our example, is $s = 1,363$. This can be used as a (virtually unbiased) estimate of the unknown standard deviation, σ, of the error term. From a substantive viewpoint, we can say that after accounting for variation in starting salaries with the grade-point average, we still have a substantial amount of variation left in starting salaries. Specifically, our estimate of the standard deviation of this unexplained variation in starting salaries is $1,363. This may be compared against the best possible result of a zero-standard deviation (i.e., no unexplained variation in starting salaries).

The coefficient of determination provides a relative measure of fit; however, instead of using the formula in Chapter 2, we propose a formula that adjusts for degrees of freedom when making inferences about the population. Specifically, for the simple linear model we use the following formula to obtain $\bar{R}^2$ (R-bar squared), or R^2 adjusted for degrees of freedom.

$$\bar{R}^2 = 1 - \frac{\Sigma(Y_i - \hat{Y}_i)^2/(n - 2)}{\Sigma(Y_i - \bar{Y})^2/(n - 1)}$$

Note that the numerator in this formula is an unbiased estimator of the variance of the errors (see Footnote 5), while the denominator is an unbiased estimator of the variance of the criterion variable. $\bar{R}^2$ provides a better and virtually unbiased estimate of the coefficient of determination in the population than does R^2.

In our example we have

$$\bar{R}^2 = 1 - \frac{7,425,962/4}{23,000,000/5} = .596$$

This tells us that if the linear model is used on the population of all observations, we expect that the predictor variable would account for almost 60 percent of the variation in the criterion variable. This estimate is lower than the (unadjusted) R^2 value of 0.677 computed in Chapter 2. $\bar{R}^2$ is always less than or at best equal to R^2. Intuitively, we can think of the difference between unadjusted and adjusted R^2 values as resulting from overfitting the data.

Given a value for R^2, $\bar{R}^2$ can be computed as follows, for a regression model with one predictor variable:

$$\bar{R}^2 = \frac{n - 1}{n - 2}(R^2) - \frac{1}{n - 2}$$

For example, suppose an analysis provided $R^2 = 0.30$ based on a sample of 7 observations using one predictor variable. Then

$$\bar{R}^2 = \frac{6}{5}(0.30) - \frac{1}{5} = 0.16.$$

In other words, with no uncertainty about the intercept and slope parameters, we *expect* to explain sixteen percent of the variation in the criterion variable in the population.

Problems

3.13 Compute the $\bar{R}^2$ for the equation in Problem 2.3, and compare the result with the unadjusted R^2 obtained in Problem 2.5.

3.14 Compute the $\bar{R}^2$ for the equation in Problem 2.4, and compare the result with the unadjusted R^2 obtained in Problem 2.6.

Confidence Intervals for Predictions

Suppose we have carried out a simple regression analysis to make specific predictions using the resulting equation. To quantify the uncertainty about such a prediction, it is often useful to construct a confidence interval. Specific predictions can be made for the conditional mean and for individual values. We distinguish between these two cases below, using the example of starting salary and grade-point average.

Conditional Mean

Sometimes we want to predict the mean value of Y based on the condition of a specific value for X, say X_o. The regression equation can be used to predict this value; thus, the best point estimate of the conditional mean is given by the equation

$$\hat{\bar{Y}}_o = a + bX_o$$

where $\hat{\bar{Y}}$ (Y-bar-hat) represents the conditional mean.

Suppose we want to predict the mean (expected) starting salary for all individuals with a grade-point average of 3.0.[7] The regression equation is

$$\hat{\bar{Y}}_o = 4{,}779 + 5{,}370X_o$$

Thus, for $X_o = 3.0$, we obtain $\hat{\bar{Y}}_o = 20{,}889$. Given a value for the predictor variable, Assumption (1) implies that $\hat{\bar{Y}}_o$ is an unbiased estimator, and Assumptions (2) and (3) imply that it is a minimum variance estimator. Together these assumptions imply that $\hat{\bar{Y}}_o$ is the best linear unbiased estimator of the conditional population mean.

Based on these three assumptions, the formula for the estimated standard error is

$$s_{\hat{\bar{Y}}_o} = s\sqrt{\frac{1}{n} + \frac{(X_o - \bar{X})^2}{\Sigma(X_i - \bar{X})^2}}$$

The fourth assumption justifies the following confidence interval:

$$\hat{\bar{Y}}_o \pm (t)(s_{\hat{\bar{Y}}_o})$$

where t is the tabulated value for $n - 2$ degrees of freedom.

With the formula for $s_{\hat{\bar{Y}}_o}$ we can see that the amount of statistical uncertainty about a prediction depends on the distance between X_o, the value upon which the prediction is conditional, and $\bar{X}$, the predictor variable's sample mean. For a prediction when $X_o = 3.0$ we have

$$s_{\hat{\bar{Y}}_o} = 1,363\sqrt{\frac{1}{6} + \frac{(3.0 - 3.3)^2}{0.54}}$$

$$\cong 787$$

And the 95-percent confidence interval is

$$\hat{\bar{Y}}_o \pm (2.78)(787) \quad \text{or}$$

$$20,889 \pm 2,188$$

Hence, the 95-percent confidence interval is 18,701 to 23,077.

Similarly, the 95-percent confidence interval for the conditional mean when $X_o = 3.8$ is obtained as follows:

$$s_{\hat{\bar{Y}}_o} = s\sqrt{\frac{1}{n} + \frac{(X_o - \bar{X})^2}{\Sigma(X_i - \bar{X})^2}}$$

$$= 1,363\sqrt{\frac{1}{6} + \frac{(3.8 - 3.3)^2}{0.54}}$$

$$\cong 1,082$$

$$\hat{\bar{Y}}_o = a + bX_o$$

$$= 4,779 + (5,370)(3.8)$$

$$= 25,185$$

At the 95-percent confidence level, as before, the tabulated t-value for 4 degrees of freedom is 2.78; thus we have

$$\hat{Y}_o \pm (t)(s_{\hat{Y}_o}) \quad \text{or}$$

$$25{,}185 \pm (2.78)(1{,}082) \quad \text{or}$$

$$25{,}185 \pm 3{,}008.$$

Hence, the 95-percent confidence interval is 22,178 to 28,193.

As indicated earlier, the statistical uncertainty about the prediction increases as the squared distance between X_o and $\bar{X}$ increases. We can verify this with the results for our two examples as summarized below:

$X_o - \bar{X}$	$s_{\hat{Y}_o}$
(3.0 − 3.3)	787
(3.8 − 3.3)	1,082

The uncertainty is smallest when $X_o = \bar{X}$ and increases the further the value for X is from this point. This phenomenon is shown graphically in Figure 3.4. As the distance between X_o and $\bar{X}$ increases, so too does the statistical uncertainty about the prediction, as indicated by the distances between the dotted curves and the straight line.

Individual Value

We often want to make predictions for one individual observation—not the conditional mean. For example, suppose you want to predict your starting salary given your grade-point average. You are not interested in quantifying the conditional mean's

Figure 3.4
Statistical uncertainty of prediction increases as distance between X_o and $\bar{X}$ increases

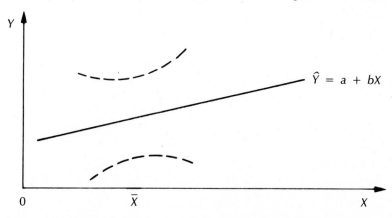

statistical uncertainty (i.e., the uncertainty about the expected starting salary conditional upon the grade-point average), but rather the uncertainty of the prediction for one individual—yourself. This uncertainty can be quantified with a formula similar to the one for the conditional mean; however, the uncertainty of individual predictions is always larger than that for conditional-mean predictions. For conditional-mean predictions, the uncertainty results from the uncertainty about the estimates of a and b. For an individual observation, this uncertainty applies also. But the error term in the equation creates additional uncertainty for the case of an individual observation. On the other hand, for a conditional mean the error is not relevant, because on the average it is zero.

Based on the assumptions, the confidence-interval formula for an individual observation is

$$\hat{Y}_o \pm (t)(s_{\hat{Y}_o})$$

where
$$s_{\hat{Y}_o} = s\sqrt{1 + \frac{1}{n} + \frac{(X_o - \bar{X})^2}{\Sigma(X_i - X)^2}}$$

Again, it is not necessary to observe the specific value of the individual observation, X_o, in the sample. Suppose your grade-point is $X_o = 3.3$. Then

$$s_{\hat{Y}_o} = 1,363\sqrt{1 + \frac{1}{6} + \frac{(3.0 - 3.3)^2}{0.54}}$$

$$\cong 1,574$$

And the 95-percent confidence interval is

$$20,889 \pm 4,376$$

Hence, the 95-percent confidence interval is 16,513 to 25,265.

On the other hand, if your grade-point average is $X_o = 3.8$,

$$s_{\hat{Y}_o} = 1,363\sqrt{1 + \frac{1}{6} + \frac{(3.8 - 3.3)^2}{0.54}}$$

$$\cong 1,740$$

And the 95-percent confidence interval is

$$\hat{Y}_o \pm (t)(s_{\hat{Y}_o}) \quad \text{or}$$

$$25,185 \pm (2.78)(1,740) \quad \text{or}$$

$$25,185 \pm 4,837$$

Hence, the lower and upper bounds are 20,348 and 30,022.

These results confirm that the uncertainty about predictions for individual observations also increases as the distance between X_o and $\bar{X}$ increases. The results for our examples are summarized below.

Distance from the sample mean $\bar{X}$	Estimated standard error of conditional mean	Estimated standard error of individual value
$(X_o - \bar{X})$	$s_{\hat{Y}_o}$	$s_{\hat{Y}_o}$
$(3.0 - 3.3)$	787	1,574
$(3.8 - 3.3)$	1,082	1,740

Problems

3.15 Using the equation obtained in Problem 2.3, obtain a 95-percent confidence interval for the conditional mean when $X_o = 2.1$ (i.e., 2100 square feet).

3.16 Using the equation obtained in Problem 2.3, obtain a 95-percent confidence interval for an individual observation when $X_o = 2.1$ (i.e., 2100 square feet).

3.17 Using the equation obtained in Problem 2.4, obtain a 90-percent confidence interval for the conditional mean when $X_o = 1.10$ (i.e., $1.10).

3.18 Using the equation obtained in Problem 2.4, obtain a 90-percent confidence interval for an individual observation when $X_o = 1.10$ (i.e., $1.10).

Variation in the Predictor Variable

To estimate a relationship between two variables it is essential that the sample contains variation in the values for the predictor variable. Without this variation it is impossible to determine how that variable may relate to the criterion variable. Further, the standard errors of the slope and the intercept, as well as the standard error of the predictions, depend on the *amount* of variation in the predictor variable.

Recall that the estimated standard error of the slope is

$$s_b = \frac{s}{\sqrt{\Sigma(X_i - \bar{X})^2}}$$

The denominator in this formula suggests that, everything else being equal, the greater the variation in the predictor variable, the smaller the estimated standard error of the slope. This amount of variation can be influenced by the sample size; as the sample size increases, the total variation in the predictor variable will tend to increase. However, the formula also suggests that for a given sample size, the estimated standard error of the slope will decrease as the variability in X increases.

For example, suppose we can choose one of two samples, each of which contains (X_i), for grade-point average. Should we choose set A or set B?

X_i(Set A)	X_i(Set B)
4.0	3.6
3.9	3.5
3.9	3.2
2.5	3.1
2.5	2.9
2.4	2.9

If we prefer the smallest possible uncertainty about the slope parameter, we should pick set A because it contains more variability in X. Specifically for set A we obtain

$$\sqrt{\Sigma(X_i - \bar{X})^2} = 1.80$$

compared with

$$\sqrt{\Sigma(X_i - X)^2} = 0.66$$

for set B. Thus, assuming that the relation between Y and X is linear, statistical significance is much more likely to be obtained for the slope coefficient if set A is used rather than set B.[8]

Statistical uncertainty associated with predictions is similarly affected by variability in X. Recall the formula for the conditional mean:

$$s_{\hat{Y}_o} = s\sqrt{\frac{1}{n} + \frac{(X_o - \bar{X})^2}{\Sigma(X_i - \bar{X})^2}}$$

As the sample size increases, $\frac{1}{n}$ becomes smaller, and $\Sigma(X_i - \bar{X})^2$ tends to increase. But also, for a given sample size, the more variability in the predictor variable, the smaller the ratio

$$\frac{(X_o - \bar{X})^2}{\Sigma(X_i - \bar{X})^2}$$

for a given X_o value. Thus, the statistical uncertainty associated with regression-based predictions becomes smaller as the total amount of variability in X increases. This relationship between statistical uncertainty and variability in X also holds for predictions of individual observations.

Note that for a prediction where $X_o = \bar{X}$, the formulas for the estimated standard error of the predictions are reduced to the following expressions:

$$s_{\hat{Y}_o} = s\sqrt{\frac{1}{n}}$$

$$s_{\hat{Y}_o} = s\sqrt{1 + \frac{1}{n}}$$

The first expression, for conditional mean predictions, shows that the standard error is identical to the standard error of a sample mean, i.e., $s/\sqrt{n}$ (ignoring the finite population correction factor). Thus, as the sample size increases, the uncertainty about the conditional mean forecast decreases. For large samples this uncertainty approaches zero; however, for individual observation forecasts, the uncertainty approaches s, the standard deviation of the residuals, as the sample size increases.

Computer Output

Although calculators can be used to obtain a regression equation for a simple linear model, it is better to use computer packages that can also display scattergrams of the data. An example of computer output for a problem involving a possible relation between dividends per share and earnings per share for AT&T is shown below.

Example

Early each year AT&T's Board of Directors determines the dividend per common share. Before this decision the company usually has available the net earnings per common share for the previous year. Conceivably then, the dividends for year t may depend on the earnings in the prior year $(t - 1)$.

The following data were collected for the AT&T company (before divestiture):

Year	Dividends paid per common share in year t (dollars)	Earnings per common share in year $(t - 1)$(dollars)
1973	2.80	4.34
1974	3.16	5.07
1975	3.40	5.26
1976	3.70	5.08
1977	4.10	5.98
1978	4.50	6.86
1979	5.00	7.74
1980	5.00	8.04

Suppose we use the following model:

$$Y_i = \alpha + \beta X_i + u_i$$

where Y_i is the dividend per common share paid in year t and

   X_i is the net earnings per common share in year $(t - 1)$.

Before applying regression analysis to the data, it is useful to graph the data. For example, the data can be plotted using Y (dividend per share) on the vertical axis and X (earnings per share) on the horizontal axis. Such a picture, often referred to as a scatterplot or scattergram, can give a feel for the data. Moreover, graphical displays help greatly in detecting unusual conditions. Chapters 6 and 7 provide a more complete discussion of the advantages associated with using plots.

The eight observations on dividend and earnings for AT&T were standardized by subtracting the sample mean and dividing the difference by the standard deviation for each variable separately.[9] These standardized values were used to produce the scatterplot shown in Figure 3.5. The formulas for the standardized variables are

$$Y_i^* = \frac{Y_i - \bar{Y}}{s_y}$$

$$X_i^* = \frac{X_i - \bar{X}}{s_x}$$

where Y_i^* is the standardized dividend (called Y-star)

   $\bar{Y}$ is the sample mean dividend (3.96)

   s_y is the standard deviation dividend (0.83)

   X_i^* is the standardized earnings (X-star)

   $\bar{X}$ is the sample mean earnings (6.05)

   s_x is the standard deviation earnings (1.36)

The plot based on the standardized dividend and earnings values makes it plain that dividends and earnings tend to be positively related: as earnings increase, dividends tend to go up. However, the relation is inexact. The plot can also be used to see whether it is reasonable to put a straight line through the data. In other words, do the data suggest a linear or a nonlinear relation? Unfortunately, with only eight data points this is difficult or impossible to determine; however, we can say that the plot is consistent with the assumption of a linear relation between Y^* and X^* and, hence, between Y and X.

We obtained the following results on the regression analysis of our example from a computer program called *Interactive Data Analysis* (*IDA*), developed at the University of Chicago.

Figure 3.5
Scatterplot of 8 standardized values of dividends vs. earnings

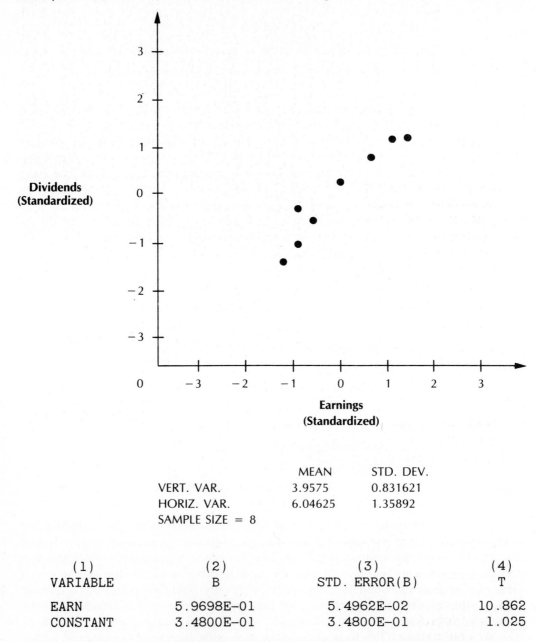

		MEAN	STD. DEV.
VERT. VAR.		3.9575	0.831621
HORIZ. VAR.		6.04625	1.35892
SAMPLE SIZE = 8			

(1)	(2)	(3)	(4)
VARIABLE	B	STD. ERROR(B)	T
EARN	5.9698E−01	5.4962E−02	10.862
CONSTANT	3.4800E−01	3.4800E−01	1.025

The second column shows the slope coefficient for earnings (EARN) and the intercept.[10] The corresponding estimated standard errors s_b and s_a are shown in the third column, and the *computed t*-values $\frac{b}{s_b}$ and $\frac{a}{s_a}$ are in the fourth column. The next piece of computer output shows summary statistics.

```
            MULTIPLE R   R-SQUARE
UNADJUSTED     0.9755      0.9516
ADJUSTED       0.9714      0.9435

STD. DEV. OF RESIDUALS = .19760
```

The summary statistics include unadjusted and adjusted R^2 values. The multiple R values are obtained by taking the square root of the corresponding R-square values. The final piece of information is the standard deviation of the residuals. (s), which is an estimate of the standard deviation of the error term (σ) in the model.

There are, of course, many software packages available for applications of regression analysis. Shown below is sample output, using our dividend and earnings example, obtained from *MINITAB*, a widely used software package.

Minitab

In the output shown below, X1 refers to earnings. The first row contains the results for the intercept, and the subsequent rows refer to the predictor variables. The standard deviation of Y about the regression line (S) is *MINITAB*'s label for the standard deviation of residuals.

```
THE REGRESSION EQUATION IS
Y = 0.348 + 0.597 X1

       COLUMN   COEFFICIENT  ST. DEV. OF COEF   T-RATIO = COEF/S.D.

         -         0.3480        0.3396             1.02
X1   EARN         0.59698        0.05496            10.86

THE ST. DEV. OF Y ABOUT REGRESSION LINE IS S = 0.1976
WITH (   8 - 2) = 6 DEGREES OF FREEDOM

R-SQUARED = 95.2 PERCENT
R-SQUARED = 94.4 PERCENT, ADJUSTED FOR D.F.
```

SUMMARY

We have introduced statistical inference in simple regression analysis. Hypothesis testing, and the construction of confidence intervals, requires four assumptions about the error term in the model. In this chapter, we have provided the formulas to perform statistical inference.

We emphasize that the justification for statistical inference about, say, a population mean is very different from the justification for inference about, say, a slope parameter. In the case of a population mean, we rely on random sampling, and this virtually guarantees the validity of appropriate statistical tests. However, in regression analysis we make assumptions about the error term that are equivalent to the implications of random sampling. These assumptions have to be valid in order to justify the statistical tests. Assessing the assumptions' validity is perhaps the most important aspect of regression analysis.

The formulas for statistical inference show the sources of uncertainty about the true parameter values. For all formulas, the variance of the error term matters. That is, the greater this variance, the greater the uncertainty (high variance means the sample observations tend to be far away from the straight line that summarizes the relation between Y and X). A second source is the amount of sample variation in X. The greater this variation, the lower the uncertainty. In addition, for predictions the uncertainty is also influenced by the distance between the value for X used for the prediction and the mean value for X in the sample. It is important to have an intuitive grasp of the reasons for these influences on the uncertainty in statistical inference.

ADDITIONAL PROBLEMS

3.19 a. Using either the *IDA* or *MINITAB* output, write the estimated equation relating dividends (DIVI) to earnings (EARN).
   b. What is the estimated increase in dividends for a one-dollar increase in earnings?
   c. Test the null hypothesis that there is no linear effect of earnings on dividends, assuming the validity of the four assumptions about the error term.

3.20 Additional computer output using *IDA* is shown below.

	Y	S.E.(FITTED)	S.E.PRED.(Y)
X'S = ?8			
	5.1239E+00	1.2811E-01	2.3550E-01
X'S = ?8.3			
	5.3029E+00	1.4221E-01	2.4346E-01
X'S = ?8.6			
	5.4820E+00	1.5679E-01	2.5225E-01

This output shows the predicted value for dividends (Y) obtained for three alternative possible values for earnings (X). An analyst might use the estimated equation to predict dividends to be declared in early 1981 based on certain expectations for 1980 earnings. For example, if earnings = $8.00, the prediction for dividends is $5.13. The third and fourth columns show the standard errors. S.E.

(FITTED) is the standard error for a conditional mean prediction, and S.E. PRED. (Y) is the standard error for the prediction of an individual observation.

a. If the earnings for 1980 are \$8.30 per common share, what is the predicted value of dividends?

b. Construct a 95-percent confidence interval for the prediction under 3.20a.

c. Suppose AT&T earnings are indeed \$8.30 in 1980, and the company declares a dividend of \$5.50 in 1981. Does this invalidate the model? Why or why not?

3.21 With reference to the previous problem, suppose that the model to be estimated is to take dividends as a function of a trend variable, that is,

$$Y_i = \alpha + \beta X_i + u_i$$

where Y_i is the dividend per common share paid in year t and

   X_i is a trend variable equal to 1 in year 1973, 2 in year 1974, etc.

The data are

Year	Dividends paid per common share in year t (dollars)	Trend
1973	2.80	1
1974	3.16	2
1975	3.40	3
1976	3.70	4
1977	4.10	5
1978	4.50	6
1979	5.00	7
1980	5.00	8

Selected computer output is shown below.

VARIABLE	B	STD. ERROR(B)	T
TREND	3.3690E−01	1.7130E−02	19.667
CONSTANT	2.4414E+00	8.6504E−02	28.223

	MULTIPLIER R	R−SQUARE
UNDAJUSTED	0.9923	0.9847
ADJUSTED	0.9910	0.9822

STD. REV. OF RESIDUALS = .111017

    a. What is the estimated increase in dividends per year?

    b. What is the predicted dividend in 1981 based on the estimated equation?

    c. Compare the equation obtained with the equation when EARN is the predictor variable (see Problem 3.19). Consider the following criteria for the purpose of comparing the two models: (1) intuitive appeal, (2) adjusted R^2, (3) unadjusted R^2, (4) computed t-ratio for the slope coefficient, (5) any other criterion (for example, how easy it is to use the equation for making forecasts). Which equation do you prefer, and why? Justify your choice of criteria.

3.22 Until the late 1960s, the military draft in the United States was handled by local draft boards. However, sending draftees to the war fields in Vietnam was a sensitive matter. Consequently, the government instituted a national draft lottery procedure to determine the draft priority status of all eligible male residents born in a given year based on the draftees' birthdates. A simple random sampling procedure was adopted to rank in order all 366 birthdates and was shown on national television.

    Each birthdate was recorded on a piece of paper and placed inside a capsule which was then placed in a glass container. After all capsules had been placed in the container, it was shaken a number of times, and one capsule was selected randomly to determine which birthdate would have priority status *one*. From the remaining capsules a second capsule was selected randomly to determine which birthdate would receive priority status *two*. This process continued until each birthdate had received a priority status.

    After the results were obtained, some people questioned the randomness of the process. In particular they noted that the capsules were created systematically, starting with January 1 and ending with December 31, and that they had been placed in the glass container in that order. Thus, the January capsules were placed on the bottom of the container, the February capsules on top of the January capsules, and so forth. If the container were insufficiently shaken, the January capsules would tend to remain on the bottom and other capsules would be closer to the top. Thus, they argued, it would be possible for the priority status to be related to the birthdate.

    To examine the possible relation between priority status and the birthdate, the median rank order of the birthdates in a month was computed. If the selection procedure were not entirely random, the median would be relatively high (low priority) for January and would slowly decrease from one month to the next. A regression analysis can be performed to examine this relation, using the median rank order of the birthdates in a given month as the criterion variable, and the month (January = 1, February = 2, etc.) as the predictor variable.

    a. Describe the null and alternative hypotheses for the slope parameter.

    b. If the procedure were entirely systematic (nonrandom), with December 31 having priority status one, December 30 having priority status two, etc., what would be the approximate slope and intercept coefficients from the regression analysis?

c. Below are the 12 data points actually obtained.

Y Median order of induction[11]	X Month
211	1
212	2
256	3
225	4
226	5
208	6
190	7
154	8
168	9
201	10
126	11
96	12

Test the null hypothesis described under 3.22a. What is the probability that the regression result would be obtained if the draft lottery were entirely random?

d. Construct a 95-percent confidence interval for the slope.

APPENDIX 3.1
COMPUTATION OF MEAN AND VARIANCE

Consider a population consisting of only six observations, that is, $N = 6$. Let the values existing in this population be $X_i = 1, 2, 3, 4, 5, 6$. The observations are assumed to be measured on at least an *interval* scale, so that it is meaningful to compute a mean. We can summarize the set of observations in this population by computing a mean and a standard deviation. Even though there are six different values in the population, each occurring once, we use the setup for a frequency distribution.

X_i	f_i	X_i^2	$X_i f_i$	$X_i^2 f_i$
1	1	1	1	1
2	1	4	2	4
3	1	9	3	9
4	1	16	4	16
5	1	25	5	25
6	1	36	6	36
	6		21	91

$$\mu = \frac{\Sigma x_i f_i}{\Sigma f_i} = \frac{21}{6} = 3.50$$

where $\Sigma = \sum_{i=1}^{k}$ and k is the number of classes.

$$\sigma^2 = \frac{\Sigma x_i^2 f_i - \dfrac{(\Sigma x_i f_i)^2}{\Sigma f_i}}{\Sigma f_i}$$

$$= \frac{91 - \dfrac{(21)^2}{6}}{6}$$

$$= \frac{17.5}{6} = 2.29$$

ENDNOTES

1. The formula for the variance assumes an infinite population size or sampling with replacement. For sampling without replacement, this formula becomes:

$$\text{Var}(\bar{Y}) = \frac{\sigma^2}{n}\left(\frac{N-1}{N-n}\right)$$

2. $\bar{X}$ will be normally distributed if X_i is normally distributed. Alternatively, we can appeal to the Central Limit Theorem to justify the assumption of a normal distribution for $\bar{X}$, if X_i does not follow a normal distribution.

3. By way of analogy, imagine a die which generates the error-term value for each observation. Rolling the die produces a result for each observation. Since each roll is independent of every other roll, each possible result is independent of every other possible result.

4. An estimator is efficient if it has the smallest variance of all unbiased estimators. Intuitively, this means that the least-squares procedure is better than other unbiased procedures for the purpose of obtaining estimates "close to" the true, but unknown, parameters.

5. Using $n - 2$ in the denominator is required to have $E(s^2) = \sigma^2$. We subtract two from n to account for the fact that two parameters, α and β, are estimated before the residuals are calculated and the standard deviation is computed.

6. Note that this test provides the same computed t-value (except for rounding errors) as the test for the slope parameter in a simple linear regression model; the test statistics are equivalent.

7. Note that the specific value does not have to be included in the sample data.

8. There are, however, other considerations which could lead us to choose set B. For example, if we want to investigate whether the assumption of linearity is reasonable, we may prefer set B, because its variability is more suitable for investigating the appropriateness of a linear model (set A has no observations between 2.5 and 3.9).

9. Due to the standardization, Y_i^* and X_i^* have an average value equal to zero and a standard deviation of one. This standardization allows the use of standard measures for the axes.

10. Note that 5.9698E − 01 equals 0.59698, and 5.4962E − 02 equals 0.054962, etc.

11. Bill Williams, *A Sampler on Sampling* (New York: John Wiley & Sons, 1978).

4

Multiple Regression Analysis with Two Predictor Variables

Chapter 2 introduced the descriptive measures associated with simple, linear regression analysis. Perhaps the most important descriptive measure is the slope coefficient, which tells us the average (expected) change in the criterion variable associated with a one-unit increase in the predictor variable. This measure only applies, however, to the sample data. Usually, we want to generalize the result to a broader context. For example, if the data pertain to the time period 1980–86, we may use the linear equation to describe the relation between the criterion and predictor variables for that time period. For the equation to be truly useful, however, we require that the result is applicable to other time periods and is not just due to chance variation. Statistical inference, properly carried out, considers the impact of chance assumptions about the error term.

Four assumptions about the error term were introduced in Chapter 3. These four assumptions allow us to justify the use of a statistical distribution along with the formulas to carry out hypothesis tests and to construct confidence intervals. However, statistical inference based on these formulas is justified only if these assumptions are valid. In this chapter we introduce one reason why the first assumption, $E(u_i) = 0$, may be violated. Substantively, we focus on invalid or misleading conclusions resulting from the use of a model with one predictor variable, when there are two relevant predictor variables. After introducing this problem with a hypothetical example, we introduce the concept of multiple, linear regression analysis (two or more predictor variables). To maintain simplicity, we do not go beyond two predictor variables in this chapter; we generalize the model to any number of predictor variables in Chapter 5.

Analogy to Random Sampling

We want to illustrate how the omission of a relevant predictor variable from regression analysis may affect the validity or generalizability of the result obtained. Before we do this, we want to provide an analogy comparing nonrandom and random sampling for estimating a population mean.

Imagine that we want to make conclusions about the mean value of a population, based on a sample or subset of the observations. If we use simple random sampling to pick a subset, then we know that every possible subset of a given size (sample size) has the same chance of being selected. Then, it follows that the expected value of the sample mean equals the population mean. Let Y_i represent the value of the ith observation in a population of seven observations:

Y_i
20
25
30
35
40
45
50

For this population, we have $\mu = 35$. If we randomly pick one observation, then

$$E(Y_i) = \mu$$

and for the model $Y_i = \mu + u_i$, we have $E(u_i) = E(Y_i - \mu) = E(Y_i) - \mu = 0$. And, simple random sampling implies that this property holds for every member of the sample, such that $E(\bar{Y}) = \mu$, for any sample size.

Now, suppose that instead of using simple random sampling, we use a procedure that systematically favors the observations with small values. Such a biased sampling scheme may produce the following set of three observations:

Y_i
20
25
30

This result is, of course, one of the possible subsets of three observations that may result from simple *random* sampling. But in our case, we know that some subsets were favored (those with small values), and for that reason we have

$$E(Y_i) \neq \mu$$

And, for the same reason, we have

$$E(\bar{Y}) \neq \mu$$

Thus, nonrandom sampling causes the sample mean to be a *biased* estimate of the population mean, and for the error term, u_i, in the model $Y_i = \mu + u_i$, we have $E(u_i) \neq 0$.

$\Sigma \hat{u}_i$ versus $E(u_i)$

We mentioned in Chapter 3 that the residuals should be used to see if there is evidence of a violation of one or more assumptions. For example, we might propose that $E(u_i) = 0$ implies that the residuals should be zero, on the average. Then if the residuals are zero, on the average, we might claim that there is no evidence of a violation of the first assumption. Unfortunately, the quantity $\Sigma \hat{u}_i$ is of no use in this regard. To see this, take the result of the biased sampling procedure discussed above. The sample mean is $\bar{Y} = 25$. Below we show the residuals ($\hat{u}_i = Y_i - \bar{Y}$), and the errors ($u_i = Y_i - \mu$). Note that the residuals represent the deviations from the sample mean, while the "errors" are the deviations from the population mean.

Criterion Variable Y_i	Residuals $Y_i - \bar{Y}$	Errors $Y_i - \mu$
20	-5	-15
25	0	-10
30	$+5$	-5
	0	-30

We see that the sum of the residuals is zero. This result always obtains. Yet, the sum of the "errors" is quite different from zero. Thus, this example illustrates why a problem with the sampling procedure cannot be detected by computing the sum or average of the residuals. Similarly, in regression analysis the validity of the first assumption about the error term in the model cannot be assessed by computing the sum or average of the residuals. Indeed, we showed in Appendix 2.3 that this sum is necessarily zero. Also, we note that in regression the validity of the first assumption is not determined by the sampling scheme but by other considerations, such as whether there are relevant variables omitted from the model. However, the nature of the difficulties of estimating a population mean caused by nonrandom sampling is analogous to the difficulties caused by omitted variables in regression.

Omitted Variables

It turns out that the first assumption, $E(u_i) = 0$, which is required for statistical inference in regression, is *usually* violated if there is one or more other relevant variables omitted from the model. By a relevant variable we mean a predictor variable that partly determines the criterion variable. Thus, using an incomplete model in regression is similar to using nonrandom sampling. More often than not, nonrandom sampling makes it impossible to generalize the results beyond the sample. Similarly, omitting relevant variables from a regression analysis makes it difficult or impossible to learn about the true effects of the predictor variable(s) included in the model.

Now, to determine whether a sampling scheme is random, we need to know the details of the sample selection procedure. In regression analysis, the problem is more complex, because there are multiple reasons for the first assumption to be violated. Having omitted variables is only one of the reasons (incorrect functional form is another one; see Chapter 6). Furthermore, even though the problem of omitted variables is easy to illustrate, and the possible consequences can be identified, it is difficult to establish in practice (1) whether potentially relevant predictor variables have been omitted, or (2) how serious the consequences are for the interpretation of results. To assess the validity of a sampling scheme, we can rely on the expertise of a sampling expert. (See Problem 3.22 for an example of an invalid sampling procedure.) In regression analysis, we need substantive expertise about the problem area to assess the completeness of the model. This expertise is required, among other things, for the identification of relevant predictor variables and for the specification of appropriate functional forms (e.g., linear vs. nonlinear).

Consequences of Omitted Variables

To illustrate the consequences of omitting a relevant predictor variable from the model, we use the following hypothetical data. Four observations are available on the demand for a product offered at different prices.

Demand (units)	Price (dollars per unit)
90	5
80	10
100	10
90	15

A plot of these data is shown in Figure 4.1.
For the simple linear model

$$Y_i = \alpha + \beta X_i + u_i$$

where Y_i is demand and
X_i is price,

the least squares result is $b = 0$. In other words, based on the four observations in the sample, price has no systematic effect on demand; this is indicated in Figure 4.1 by the horizontal line for an average of 90 units. With price as the only predictor variable, we conclude that demand does not depend on price.

Figure 4.1
Scattergram of demand and price

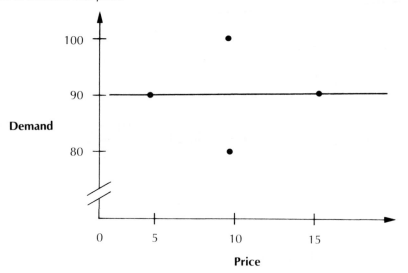

However, another predictor variable may also affect demand. For example, suppose that the product has been advertised at the different expenditure levels shown below.

Y_i Demand (units)	X_{1i} Price (dollars)	X_{2i} Advertising (dollars)
90	5	100
80	10	100
100	10	300
90	15	300

If demand depends on price as well as on advertising, we should estimate the effect of price on demand, holding advertising expenditure *constant*, (and, the effect of advertising on demand, holding price constant). If both price and advertising affect demand, then in general a valid indication of the effect of price on demand cannot be obtained without explicitly considering advertising's effect on demand at the same time. To do this we can expand our model by including advertising as a predictor variable:

$$Y_i = \alpha + \beta_1 X_{1i} + \beta_2 X_{2i} + u_i \qquad (4.1)$$

where Y_i is demand

X_{1i} is price

X_{2i} is advertising

In equation (4.1), β_1 indicates the true linear effect of X_1 on Y, holding X_2 constant, and β_2 indicates the true linear effect of X_2 on Y, holding X_1 constant. Note that in this model, we assume that the effect of price is indicated by β_1, and that this parameter does not depend on the amount of advertising. That is, we do not allow for an interaction effect between price and advertising. To estimate these parameters we have to use a procedure which allows us to estimate the effect on demand of one variable by holding the other constant.

We can do this quite easily for our example. For example, we can measure the effect of price on demand when the advertising expenditures are $100. Since there are only two data points, a straight line fits these two observations perfectly. The slope coefficient for X_1, price, can be found by investigating only the two observations for which $X_2 = \$100$:

$$b_1 = \frac{\triangle \text{ Demand}}{\triangle \text{ Price}} = \frac{90 - 80}{5 - 10} = -2$$

Note that the hypothetical data have been conveniently constructed so that the answer does not change if the slope coefficient is computed for $X_2 = \$300$. Thus, when the advertising expenditures are $300,

$$b_1 = \frac{\triangle \text{ Demand}}{\triangle \text{ Price}} = \frac{100 - 90}{10 - 15} = -2$$

This confirms that the price effect does not depend on the amount of advertising (i.e., there is no interaction effect).

Summarizing the results so far, when we used a simple regression model, we obtained a slope coefficient for the price variable equal to zero; however, when we used a multiple regression model and held advertising constant, we obtained a slope coefficient for price of -2. Thus, including additional predictor variables may change the result for a given predictor variable. We must then ask of these which is the more reasonable result. To answer this question we require substantive knowledge about the problem.

If we believe that demand is affected by price as well as by advertising, then both predictor variables should appear in the model. Hence, multiple regression is the better model. The simple regression result is erroneous because the *measured* effect of price is confounded by the variation in advertising. From the four data points we can see that the lower advertising budget ($100) was used when the price was relatively low ($5 and $10), and the higher advertising budget ($300) was used with a relatively high price ($10 and $15). In other words, there is a positive correlation between the two predictor variables.

Given that historical—as opposed to experimental—data are commonly used in regression analysis, there is nothing that can be done about correlations between predictor variables. When such correlations exist, simple regression analysis (e.g., advertising omitted) provides biased results. In our example, multiple regression produces the finding that the slope coefficient for price equals -2 (i.e., for a one-dollar increase in price, demand declines by two units); simple regression misleadingly suggests that the slope coefficient equals 0.

A Main-Effects, Linear Multiple Regression Model

In our example, we found the slope coefficient for price to be -2 at both advertising levels. If that were not the case, we would have to consider the possibility that the effect of price on the demand depends on the amount of advertising expenditures. Such a complication is called an interaction effect and is different from the existence of a correlation between the two predictor variables. In our example there is no interaction effect, but there is positive correlation between price and advertising. Conversely, it is possible for an interaction effect to exist when the correlation coefficient is zero. And it is possible that there is no correlation and no interaction or that both exist at the same time. We treat interaction effects in more detail in Chapter 5.

If we adopt equation (4.1) as the model for demand as a function of price and advertising, we are implicitly assuming (1) main effects only for each predictor variable, and (2) linear effects for each predictor. To allow only main effects means that we are allowing for no interactions. We can see this by using the estimated equation for our example. We already found that $b_1 = -2$. The effect of advertising, holding price constant, can be obtained in a similar manner. Specifically, there are two observations for which price does not vary ($X_{1i} = \$10$). Then,

$$b_2 = \frac{\triangle \text{ Demand}}{\triangle \text{ Advertising}} = \frac{100 - 80}{300 - 100} = 0.1$$

Thus, $b_2 = 0.1$, the estimated increase in demand for a one-dollar increase in advertising, holding price constant. Finally, the intercept is the estimated demand when both price and advertising are zero. By extrapolation we find the intercept, $a = 90$. The equation is

$$\hat{Y}_i = 90 - 2X_{1i} + 0.1X_{2i}$$

Problems

4.1 Using the estimated equation for demand as a function of price and advertising, compute the four fitted ($\hat{Y}$) values for the sample data, and verify that the model provides a perfect fit for the data.

4.2 Show, by example, that the estimated equation does not include any interaction effects. Specifically, assume that $X_2 = \$200$, and compute the difference in demand for $X_1 = \$10$ and $\$11$. Then, assume that $X_2 = \$400$, and again compute the difference in demand for X_1 = \$10 and \$11. The fact that the difference in demand for the two price levels is the same at both advertising levels illustrates the implicit assumption of no interaction effects.

A Graphical Illustration

We can also illustrate the basic concept of a main-effects multiple regression graphically. To do this, we focus on the relation between demand and price, holding advertising constant. This relation is depicted in Figure 4.2. The graph shows two *parallel* straight lines, with each line representing the effect of price on demand at a given advertising level. The existence of parallel lines makes it clear that advertising shifts the position of the straight line, without changing its slope (i.e., without interaction).

We treat the problem of omitted variables formally in the section on specification error later in this chapter. We do, however, want to emphasize that the consequences of omitted variables are most serious when the omitted variables have considerable effect on the criterion variable *and* are correlated with the included predictor variable. Thus, it is especially important to hold other predictor variables constant in those cases. However, the data are usually not as convenient as in our hypothetical data set. Nevertheless, this example is very useful for a conceptual explanation of separately estimating the effects of two predictor variables, and it effectively illustrates the idea of multiple regression analysis. We now discuss the least-squares method for a two-predictor variable model, which formalizes the approach used.

Figure 4.2
Effect of price on demand, holding advertising constant

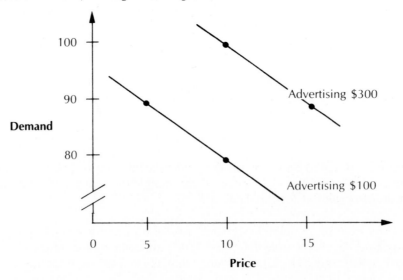

The Least-Squares Method

The model is

$$Y_i = \alpha + \beta_1 X_{1i} + \beta_2 X_{2i} + u_i$$

Given a set of data on Y, X_1, and X_2, we want to obtain estimates of the parameters α, β_1, and β_2. As in Chapter 2, we can obtain formulas for these estimates a, b_1, and b_2 by minimizing the sum of the squared residuals. That is, let

$$\hat{Y}_i = a + b_1 X_{1i} + b_2 X_{2i} \quad \text{and}$$

$$\hat{u}_i = Y_i - \hat{Y}_i$$

Then by stating that we want to have an equation which minimizes $\Sigma \hat{u}_i^2$, we obtain formulas for a, b_1, and b_2 (see Appendix 4.1). These formulas are much more complex than the formulas used for simple, linear regression analysis. To minimize the space required for the formulas, we express all variables in deviation form. That is, let

$$y_i = Y_i - \bar{Y}$$

$$x_{1i} = X_{1i} - \bar{X}_1$$

$$x_{2i} = X_{2i} - \bar{X}_2$$

Then,

$$b_1 = \frac{(\Sigma y_i x_{1i})(\Sigma x_{2i}^2) - (\Sigma y_i x_{2i})(\Sigma x_{1i} x_{2i})}{(\Sigma x_{1i}^2)(\Sigma x_{2i}^2) - (\Sigma x_{1i} x_{2i})^2}$$

$$b_2 = \frac{(\Sigma y_i x_{2i})(\Sigma x_{1i}^2) - (\Sigma y_i x_{1i})(\Sigma x_{1i} x_{2i})}{(\Sigma x_{1i}^2)(\Sigma x_{2i}^2) - (\Sigma x_{1i} x_{2i})^2}$$

For the intercept the formula is simply

$$a = \bar{Y} - b_1 \bar{X}_1 - b_2 \bar{X}_2$$

To summarize, the formula for b_1 provides the estimated linear effect of X_1 on Y, holding X_2 constant; similarly, the formula for b_2 provides the estimated linear effect of X_2 on Y, holding X_1 constant. Holding X_2 constant to obtain a valid estimate of the effect of X_1 on Y is especially important if X_2 is correlated with X_1 *and* X_2 affects Y. If X_1 and X_2 are not correlated, it can be shown that there is no need to hold X_2 constant. The formula for the intercept gives the estimated value for the criterion variable when both X_1 and X_2 are zero.

Zero Correlation between X_1 and X_2

If X_1 and X_2 are not correlated, then the product $\Sigma x_{1i} x_{2i} = \Sigma(X_{1i} - \bar{X}_1)(X_{2i} - \bar{X}_2)$ is zero. In that case, the formula for b_1 simplifies to

$$b_1 = \frac{(\Sigma y_i x_{1i})(\Sigma x_{2i}^2)}{(\Sigma x_{1i}^2)(\Sigma x_{2i}^2)} = \frac{\Sigma y_i x_{1i}}{\Sigma x_{1i}^2}$$

It is easy to see that this is exactly the same formula as we would use in a simple regression analysis for Y as a linear function of X_1. Thus, it is the second part of the b_1 formula that holds the other predictor variable constant. This second part (the part that disappears when X_1 and X_2 are not correlated in the sample) captures the correlation between Y and X_2 as well as the correlation between X_1 and X_2.

Again, the greater the degree of correlation between X_1 and X_2, the more important it is to use multiple regression analysis, if both X_1 and X_2 are relevant predictor variables. For example, the difference in the computed slope coefficient for a predictor variable from a simple, as opposed to a multiple, regression analysis increases as the correlation between the two predictor variables increases. Thus, even if we are only interested in the effect of X_1, we must also include X_2 in the regression. However, if X_1 and X_2 are not correlated in the sample, but both X_1 and X_2 are relevant predictor variables, it turns out to be advantageous to include X_2 in the model. This advantage shows up in the estimated standard error of the slope coefficient, as we see later.

An Application

Consider the following data set relevant to the problem of demand as a function of price and advertising.

Y_i Demand (units)	X_{1i} Price (dollars)	X_{2i} Advertising (dollars)
50	11	400
90	13	1200
70	12	600
50	14	800
90	10	800

Many software packages provide a correlation matrix for the variables first. We follow this convention here and interpret the entries in the matrix shown below.

Correlation matrix

	Demand	Price	Advertising
Demand	1.00		
Price	−0.32	1.00	
Advertising	0.67	0.43	1.00

Each correlation coefficient describes the degree of linear association between two variables for the sample data. A coefficient is computed by using only the data relevant to the pair of variables in question. For example, the correlation of −0.32 for demand and price is obtained with the correlation coefficient formula shown in Chapter 2. These coefficients are called *simple* correlation coefficients, because no other variables have been held constant in the computations. For that reason, the sign of the slope coefficient for price in a multiple regression may differ from the sign of the correlation coefficient between demand and price.

We can use the correlation matrix to determine the degree of linear association between the predictor variables. The positive correlation of 0.43 between price and advertising suggests that these variables tend somewhat to move together. It also tells us that a simple regression of demand as a function of price will give us a different result than a multiple regression will. Given that we believe that both price and advertising are relevant predictor variables, we use multiple regression. In practice, a computer is used to obtain results for a multiple regression, but we illustrate the use of the formulas for this sample of five observations.

We first compute the sample means: $\bar{Y} = 70$, $\bar{X}_1 = 12$, and $\bar{X}_2 = 760$. Then, we complete the computations using the observations expressed in deviation form:

y_i	x_{1i}	x_{2i}	$y_i x_{1i}$	$y_i x_{2i}$	$x_{1i} x_{2i}$	x_{1i}^2	x_{2i}^2
−20	−1	−360	20	7200	360	1	129,600
20	1	440	20	8800	440	1	193,600
0	0	−160	0	0	0	0	25,600
−20	2	40	−40	−800	80	4	1,600
20	−2	40	−40	800	−80	4	1,600
0	0	0	−40	16,000	800	10	352,000

$$b_1 = \frac{(-40)(352,000) - (16,000)(800)}{(10)(352,000) - (800)^2} = -9.3$$

$$b_2 = \frac{(16,000)(10) - (-40)(800)}{(10)(352,000) - (800)^2} = 0.067$$

$$a = 70 - (-9.3)(12) - (0.067)(760) = 131$$

The estimated equation is therefore

$$\hat{Y}_i = 131 - 9.3X_{1i} + 0.067X_{2i}$$

We can also see directly from the numbers in the formula for b_1 that the price coefficient would be -4.0, if advertising is not held constant. That is, in simple regression analysis, the formula for the price coefficient is

$$b_1 = \frac{\Sigma y_i x_{1i}}{\Sigma x_{1i}^2} = \frac{-40}{10} = -4$$

Amusing that both price and advertising are relevant, the simple regression result underestimates the price effect. This systematic effect can be understood as follows. Advertising has a positive correlation with price. When the price increases, demand is expected to fall. However, if advertising tends to increase at the same time (positively correlated with price), demand will not decrease as much as when advertising stays constant. For that reason, the effect of price based on a simple regression will be understated in this case.

We conclude that multiple regression has the appealing property that it provides separate slope coefficients for each predictor variable, holding the other predictor variable(s) constant. It is important that this is done in order to obtain valid estimates of the effects we are interested in. Omitting a relevant predictor variable systematically biases the effect of the predictor variable of interest, with the extent of the bias dependent upon the correlation between the predictor variables. If a relevant predictor variable is omitted because of a lack of data, we can speculate about its effect on the computed slope coefficient based on knowledge or expectations about the sign, and degree of simple correlation between the included and omitted predictor variables.

For example, if there are no data available on advertising expenditures, we have to ask ourselves two questions: (1) What is the nature of the effect of advertising on demand (answer: generally, positive)? and (2) What is the nature of the correlation between price and advertising (answer: positive, if management believes that higher prices need to be supported with higher advertising expenditures)? If we are confident about such answers, we know the nature of the bias in the simple regression price coefficient.

Goodness of Fit

The unadjusted R^2 for the equation equals 0.90. This measure can be computed in several ways. One formula provides the result based directly on the three simple correlation coefficients (the formula is shown in Appendix 4.2 of this chapter).

Generally, it is easier to use the standard formula

$$R^2 = 1 - \frac{\Sigma(Y_i - \hat{Y}_i)^2}{\Sigma(Y_i - \bar{Y})^2}$$

For a model with two predictor variables, adjusted R^2 is obtained by taking

$$\bar{R}^2 = 1 - \frac{\Sigma(Y_i - \hat{Y}_i)^2/(n - 3)}{\Sigma(Y_i - \bar{Y})^2/(n - 1)}$$

The relevant computations are shown below.

Y_i	$\hat{Y}_i$	$Y_i - \hat{Y}_i$	$(Y_i - \hat{Y}_i)^2$	$(Y_i - \bar{Y})^2$
50	55.3	−5.3	28.4	400
90	90.0	0	0	400
70	59.3	10.7	113.8	0
50	54.0	−4.0	16.0	400
90	91.3	−1.3	1.8	400
			160.0	1600

Thus, unadjusted R^2 equals

$$R^2 = 1 - \frac{160}{1600} = 0.90$$

And adjusted R^2 equals

$$R^2 = 1 - \frac{160/2}{1600/4} = 0.80$$

Conceptually, in determining the unexplained variation we have lost three degrees of freedom, because we computed b_1, b_2, and a before calculating the residuals. As a result there are only $(n - 3)$ or 2 degrees of freedom left for the numerator in the adjusted R^2 formula. Recall from Chapter 3 that adjusted R^2 is a better estimate of the model's explanatory power. With more data on the same problem, and a similar range of variation on the two predictor variables, we expect the explanatory power of the model to be 80 percent. Specifically, adjusted R^2 provides an estimate of the explanatory power of the model if another data set were available on the same problem, and the parameters were estimated based on that data set.

Problems

For the following problems use any of the available computer packages for regression analysis.

4.3 Enter the following data for three variables into a computer file.

Y Demand (thousands of units)	X_1 Price per unit (dollars)	X_2 Advertising (thousands of dollars)
445	10	25
385	14	20
465	10	30
425	12	25
450	12	30
455	8	20
425	14	30
495	8	30
460	8	25
415	12	20
430	10	20
420	14	25

a. Obtain the simple correlation matrix for all pairs of the three variables. Interpret the entries.
b. Obtain scattergrams for Y versus X_1, and Y versus X_2. Interpret the plots.
c. Obtain the coefficients for the linear model $\hat{Y}_i = a + bX_{1i}$, as well as for the model $\hat{Y}_i = b_0 + b_1X_{1i} + b_2X_{2i}$. Compare the slope coefficients for X_1 for the two models. Are they the same or different? Why?
d. Obtain unadjusted and adjusted R^2 values for the multiple regression analysis. These will be used in subsequent problems.

4.4 Enter the following data for three variables on twelve homes into a computer file.

Y List price ($000)	X_1 Number of Square Feet	X_2 Number of Bedrooms
117	1,000	2
187	1,800	4
232	2,400	4
133	1,600	3
115	1,400	2
213	2,200	4

Y List price ($000)	X_1 Number of Square Feet	X_2 Number of Bedrooms
175	1,800	2
154	1,900	3
151	1,400	3
182	2,200	3
134	1,600	2
185	2,300	4

a. Obtain the simple correlation matrix for all pairs of the three variables. Interpret the entries.

b. Obtain scattergrams for *Y* versus X_1, and *Y* versus X_2. Interpret the plots.

c. Obtain the coefficients for the linear model $\hat{Y}_i = a + bX_{1i}$, as well as for the model $\hat{Y}_i = b_0 + b_1X_{1i} + b_2X_{2i}$. Compare the slope coefficients for X_1 for the two models. Are they the same or different? Why?

d. Obtain unadjusted and adjusted R^2 values for the multiple regression analysis. These will be used in subsequent problems.

A Test of the Equation as a Whole

To determine the statistical significance of a multiple regression analysis, we evaluate the equation as a whole. We continue to assume a model with two predictor variables; that is,

$$Y_i = \alpha + \beta_1X_{1i} + \beta_2X_{2i} + u_i$$

The estimated equation is

$$\hat{Y}_i = a + b_1X_{1i} + b_2X_{2i}$$

Now, imagine that we are primarily interested in the slope coefficient for X_1; that is, we may want to construct a confidence interval around b_1 or test a hypothesis about the value of the parameter β_1. Still, it is necessary to first test the statistical significance of the equation for several reasons.

1. With two predictor variables in the equation, there are two opportunities for obtaining statistical significance for a slope coefficient. In essence, the type I error for statistical inference about an individual slope parameter becomes greater than the level assumed. As the number of predictor variables included in the model increases, the probability that at least one of the variables has a statistically significant coefficient becomes large, even if none of the predictor variables is truly related to the criterion variable in the population. Therefore, we first need to determine whether the equation as a whole is better than would result from chance variation.

2. If the predictor variables are substantially correlated, it becomes difficult to obtain reliable *separate* estimates of the slope coefficients. Suppose that each predictor variable alone can explain a substantial amount of variation in the criterion variable, and that the two predictor variables together explain only slightly more than each separately. Then the equation as a whole may be useful for prediction, even though it is difficult to estimate the separate influence of each predictor variable with an acceptable degree of reliability (i.e., low standard error). This is commonly referred to as a problem of multicollinearity. We discuss this aspect shortly and show how an equation can be statistically significant without either of two slope coefficients having significant *t*-ratios.

A Statistical Test of the Overall Multiple Regression

Under the null hypothesis, no effect for any of the predictor variables is assumed; thus, we have

H_0: $\beta_1 = \beta_2 = 0$

H_A: at least one predictor variable affects the criterion variable

This hypothesis can be tested indirectly based on the R^2 value. Specifically, given the assumptions about the error term in the model we can show that

$$\frac{R^2/(m)}{(1 - R^2)/(n - m - 1)}$$

follows the F distribution, with m and $(n - m - 1)$ degrees of freedom. That is, under the null hypothesis, the above ratio is distributed according to the F-distribution with m degrees of freedom in the numerator and $(n - m - 1)$ degrees of freedom in the denominator. Also,

R^2 is the unadjusted R^2 value,

m is the number of slope parameters in the model, and

n is the number of observations.

Note that the computed value of F is based on the ratio of explained variation over unexplained variation. In the numerator we divide by the degrees of freedom *used* to estimate the unknown β values, whereas in the denominator we divide by the degrees of freedom *left* after estimating α and the unknown β values.

In our demand example we have

$$R^2 = 0.90$$

$$m = 2$$

$$n = 5$$

Then

$$F = \frac{0.90/2}{0.10/2} = 9$$

We compare this computed value to the tabulated value for $m = 2$ degrees of freedom in the numerator and $n - m - 1 = 2$ degrees of freedom in the denominator. Based on a table for the F-distribution, $F_{0.05,2,2} = 19$. That is, for 2 and 2 degrees of freedom, there is only a five-percent chance of obtaining a computed value of 19 or higher if the null hypothesis ($\beta_1 = \beta_2 = 0$) is correct. Since the computed value is only 9, we cannot reject the null hypothesis at the five-percent level. However, if a type I error probability of ten percent were chosen, the result would be just statistically significant, since $F_{0.10,2,2} = 9$.

If the model as a whole is statistically reliable, the statistical significance of individual coefficients in the model can then be determined. Again, the assumptions about the error term have to be satisfied for the procedure discussed below to be meaningful.

Problems

4.5 Determine whether the multiple regression equation obtained in Problem 4.3 is statistically significant at the five-percent level.

4.6 Determine whether the multiple regression equation obtained in Problem 4.4 is statistically significant at the five-percent level.

A Statistical Test of One Selected Slope Parameter

Suppose we now want to determine the statistical significance of the estimated linear effect of price (X_1) on demand. If we expect the effect to be negative before collecting data, then we can use a one-tailed test as follows:

$$H_0: \beta_1 \geq 0$$

$$H_A: \beta_1 < 0$$

The test statistic is

$$t = \frac{b_1 - \beta_1}{s_{b_1}}$$

where b_1 is the estimated effect of price,

β_1 is the value under H_0 for which it is most difficult to reject the null hypothesis[1] (i.e., for $\beta_1 = 0$), and

s_{b_1} is the estimated standard error of b_1.

Note that the test statistic is very similar to the one used for the slope in a simple linear model; however, the formulas for b_1 and s_{b_1} are considerably more involved than the corresponding formulas used for simple linear regression. For the estimated standard error we have

$$s_{b_1} = \frac{s}{\sqrt{\dfrac{\Sigma x_{1i}^2 \Sigma x_{2i}^2 - (\Sigma x_{1i} x_{2i})^2}{\Sigma x_{2i}^2}}}$$

where

$$s = \sqrt{\frac{\Sigma(Y_i - \hat{Y}_i)^2}{n - 3}}$$

$$\hat{Y}_i = a + b_1 X_{1i} + b_2 X_{2i}$$

$$x_{1i} = X_{1i} - \bar{X}_1$$

$$x_{2i} = X_{2i} - \bar{X}_2$$

Obviously, these computations are cumbersome; consequently, the computer is used to obtain the coefficients, estimated standard errors, and so forth.

In the example we obtained $b_1 = -9.3$, and, based on computer calculations, $s_{b_1} = 3.1$. Thus, the computed value for the statistic is

$$t = \frac{-9.3}{3.1} = -3$$

This computed value can be compared with the tabulated value for $n - m - 1 = 2$ degrees of freedom. According to the table, $t_{0.05,\ 2df} = 2.92$ for a one-tailed test. Since the computed (absolute) value exceeds the tabulated value, we can reject the null hypothesis. The estimated price effect on demand is statistically significant at the five-percent level.

Most computer programs automatically provide standard errors and t-values; however, there is no guarantee that the procedure for determining the statistical significance of a result is appropriate. In other words, the computer takes the four error-term assumptions as given. In practice, we have to investigate whether these assumptions are satisfied (see Chapter 7). If the error-term assumptions are valid, then we can carry out the test shown above, as well as similar tests about the unknown values of β_2 and α in the model, or we can construct confidence intervals.

Problems

4.7 Construct a 95-percent confidence interval for the effect of price on demand, using the multiple regression result in Problem 4.3.

4.8 Test the null hypothesis that advertising does not affect the demand, that is, $\beta_2 = 0$, in the multiple regression in Problem 4.3.

4.9 Construct a 98-percent confidence interval for the effect of square feet on list price, using the multiple regression result in Problem 4.4.

4.10 Test the null hypothesis that the number of bedrooms does not affect the list price, that is, $\beta_2 = 0$, in the multiple regression in Problem 4.4.

Adding X_2 when X_1 and X_2 are Not Correlated

We have established that the consequence of omitting a relevant predictor variable on the validity of the estimated effect of the included variable becomes more severe as the correlation between X_1 and X_2 increases. Further, there are no consequences for the *validity* of the estimate at all, if X_1 and X_2 are uncorrelated. Nevertheless, we can increase the *reliability* of the estimate by adding a relevant predictor variable to the model. Remember that by validity we mean that $E(b_1) = \beta_1$ for the effect of X_1. Reliability, on the other hand, refers to the uncertainty or estimated standard error of the slope. The lower this standard error, the more reliable the estimated effect.

To see the effect of adding a relevant predictor variable (X_2) when this variable is not correlated with the predictor variable of interest (X_1), we use the formula for the estimated standard error of b_1.

$$s_{b_1} = \frac{s}{\sqrt{\dfrac{\Sigma x_{1i}^2 \Sigma x_{2i}^2 - (\Sigma x_{1i} x_{2i})^2}{\Sigma x_{2i}^2}}}$$

If X_1 and X_2 are uncorrelated, $\Sigma x_{1i} x_{2i} = 0$. The formula then reduces to

$$s_{b_1} = \frac{s}{\sqrt{\Sigma x_{1i}^2}}$$

This is virtually the same formula as the formula for the standard error of the slope from a simple regression analysis. However, although the denominator is identical to that in the earlier equation, the numerator is not. If the second predictor variable is relevant, then the unexplained variation is reduced. Consequently, the standard deviation of the residuals may be reduced as well. More precisely, s_{b_1} (the formula shown above, when X_1 and X_2 are uncorrelated) will be smaller than s_b (the formula for a simple, linear regression), if the additional variation explained by X_2 is greater than the loss of an additional degree of freedom.

The Sampling Analogy, Once More

We stress again the similarity between the model specification in regression analysis and the sampling procedure for estimating the mean value of a population. Simple random sampling provides an *unbiased* estimate of the population mean. However, it does not necessarily give an *efficient* estimate. That is, there are often other sampling procedures that are superior to simple random sampling (lower standard error of the mean with the same sample size, or lower cost of sampling for the same standard error of the mean). For example, the population may be stratified prior to sampling; a stratified random sample would provide an unbiased estimate of the population mean, often with smaller uncertainty (lower standard error) than a simple random sample of the same size. Specifically if the variability in the observations differs between strata or subgroups, stratified random sampling is more efficient than simple random sampling. We have summarized the most relevant aspects of sampling for the estimation of a population mean and the specification of a second predictor variable in regression analysis in Table 4.1.

Table 4.1
Similarities between sampling and regression analysis

	Estimation of a population mean	Estimation of a linear relation
Model	$Y_i = \mu + u_i$	$Y_i = \alpha + \beta_1 X_{1i} + \beta_2 X_{2i} + u_i$
Parameter of primary interest	μ	β_1
Estimate based on sample data	$\bar{Y}$	b_1
Desirable properties		
1. unbiasedness	$E(\bar{Y}) = \mu$	$E(b_1) = \beta_1$
2. efficiency	$Var(\bar{Y})$ smallest of all unbiased estimators	$Var(b_1)$ smallest of all unbiased estimators
Example of violation of 1 (unbiasedness)	Nonrandom sampling	Omission of X_2 when it is correlated with X_1, and X_2 is relevant
Example of violation of 2 (efficiency)	Simple as opposed to stratified random sampling if $Var(u_i) \neq \sigma^2$ for all i (or, one or more subsets have higher variability than other subsets)	Omission of X_2 when it is not correlated with X_1, and X_2 is relevant

In statistics, we are interested in unbiased and efficient estimators. As we have discussed, omitting a relevant predictor variable tends to result in biased and inefficient estimators. Bias results if the omitted variable is correlated with the included variable. Inefficiency results especially if the omitted variable is not correlated with the included variable.

Statistical Reliability of Predictions

The equation with two predictor variables can also be used for predictions. If these predictions are used, it would be worthwhile to assess statistical reliability. Again, confidence intervals constructed around a predicted value are valid only if the assumptions underlying the procedure are met. Computer programs can be used to obtain the predictions, for given values of the predictor variables, and their standard errors. We used the *IDA* program to obtain predictions for some selected values shown below in the demand example discussed earlier.[2]

Suppose we are considering two alternative strategies for influencing demand. One strategy consists of a high price and high advertising expenditures; the other uses a low price and little advertising. The specifics of the two strategies are

	Price per unit (dollars)	Advertising (dollars)
Strategy A	14	1200
Strategy B	10	400

Based on the computer program we obtain

	Predicted demand	Estimated standard error[3] of individual value
Strategy A	80.7	12.2
Strategy B	64.7	11.8

For a 90-percent confidence interval we use the tabulated value of t (with $n - 3 = 2$ degrees of freedom) equal to 2.920. Thus, strategy A has a predicted demand of 80.7 with a lower bound of $80.7 - (2.92)(12.2) = 45.1$ and an upper bound of $80.7 + (2.92)(12.2) = 116.3$. Similarly, strategy B's predicted demand is 64.7 with a lower bound of 30.2 and an upper bound of 99.2.

Remember, these calculations consider only statistical reliability. The model may not capture the essence of the market's reactions to the firm's strategies. For example, the influence of price and advertising may depend on economic climate. If this climate changes, the effects of these marketing variables may change too. Furthermore, these effects may depend also on the number of competitors in the industry, the product's quality, and many other factors. Statistical analysis does not address these questions. A user of statistical analysis *always* has to consider the model's validity. Indeed, the marketplace's complexity is far greater than this model suggests, and statistical inference is justified only if the model is a reasonably valid representation of the problem (see also Chapter 7).

Problems

4.11 Construct a 95-percent confidence interval for the demand if the price equals $10 and advertising expenditures are $25,000 (i.e., $X_1 = 10$, and $X_2 = 25$), using the multiple regression in Problem 4.3.

4.12 Construct a 95-percent confidence interval for the list price of a house that has 2,500 square feet and 4 bedrooms (i.e., $X_1 = 2,500$, and $X_2 = 4$), using the multiple regression in Problem 4.4.

Computer Output

Printed below are various pieces of computer output showing the results used in this example.[4] With the command "COEF" we obtained regression coefficients[5] (slope coefficients and intercept), standard errors, and computed *t*-values (under a null hypothesis of no effect). These results differ somewhat from the computations shown earlier in this chapter because of rounding.

```
> COEF

VARIABLE              B              STD. ERROR(B)           T

PRICE            -.9333E+01          3.1269E+00         -2.985
ADVERT           6.6667E-02          1.6667E-02          4.000
CONSTANT         1.3670E+02                             3.612
```

The command "SUMM" provides us with the unadjusted and adjusted R^2 values, their square roots, and the standard deviation of residuals.

```
> SUMM

                 MULTIPLE R    R-SQUARE

UNADJUSTED        0.9487        0.9000
ADJUSTED          0.8944        0.8000

STD. DEV. OF RESIDUALS = .894426
```

Next we obtained an analysis of variance with the command "ANOV." The item of direct interest here is the computed F-value shown in the last column. This value is used to determine the overall equation's statistical significance.

```
> ANOV
```

SOURCE	SS	DF	MS	F
REGRESSION	1.44000E+01	2	7.20000E+00	9.00
RESIDUALS	1.60000E+00	2	7.99998E-01	
TOTAL	1.60000E+01	4	4.00000E+00	

The analysis of variance includes the computed F-value as follows for a model with two predictor variables:

$$F = \frac{\Sigma(\hat{Y}_i - \bar{Y})^2/2}{\Sigma(Y_i - \hat{Y}_i)^2/(n-3)} = \frac{SS \text{ regression}/2}{SS \text{ residuals}/(n-3)}$$

where SS is sum of squares.

This is equal to

$$F = \frac{R^2/2}{(1 - R^2)/(n-3)} = \frac{0.90/2}{0.10/2} = 9.00$$

Finally, using the command "SEPR" we obtain predictions for the two strategies. The X's are the values for the predictor variables we want predictions for (e.g., strategy A uses a \$14 price and \$1200 in advertising expenditures). The first column (Y) shows the predicted demand, the second column gives the standard error for the prediction of a conditional mean (S.E.FITTED), and the third column shows the standard error for the prediction of an individual value (S.E.PRED (Y)).

```
> SEPR
```

	Y	S.E.FITTED	S.E.PRED(Y)
X'S = ?14,1200	8.0667E+01	8.3533E+00	1.2238E+01
X'S = ?10,400	6.4667E+01	7.6884E+00	1.1795E+01

Multicollinearity

As we mentioned at the beginning of this chapter, the results from a regression analysis are likely to be misleading if one or more relevant predictor variables have been omitted from the model. Hence, all predictor variables which affect the criterion variable should be included in the regression equation. Including all relevant variables and properly specifying the nature of the effects (e.g., linear versus nonlinear, see Chapter 6) ensure that the estimated effects are unbiased. We also mentioned that a correlation between two relevant predictor variables makes it especially important that both predictor variables are included. Ironically, however, the higher this correlation, the more difficult it is to obtain reliable estimates of the separate effects.[6] The estimates are still valid (if the model is well-specified), but may encounter unacceptably high standard errors for the slope coefficients. We show below how, everything else being held constant, as the correlation between the two predictor variables increases, the estimated standard errors increase as well.

The degree of correlation between predictor variables is commonly referred to as collinearity, or multicollinearity. We demonstrate the problem of collinearity below, using an example. In particular, we want to demonstrate that the estimated standard errors may become so high that it is impossible to reject the null hypothesis of no effect for each predictor variable. Yet, the multiple regression equation can have statistical significance, from which we conclude that at least one slope parameter is different from zero.

As before, we have data on three variables—one criterion variable and two predictor variables. From sample data on these three variables, we want to compute b_1, b_2, and a. Thus there are three equations and three unknown values (see Appendix 4.1). A unique solution can be found only if the three equations are linearly independent. This condition of linear independence is violated if the predictor variables, X_1 and X_2, are perfectly correlated, either positively or negatively. In that case one predictor variable must be excluded from the model, in order to get a solution. This is an example of extreme multicollinearity.

In practice it is more common to find that two predictor variables are highly, but not perfectly, correlated. To obtain a valid estimate of one predictor variable's influence on the criterion variable, the other predictor variable must be held constant; however, the higher the correlation between two predictor variables, the more difficult this is to do. Under extreme multicollinearity it is impossible.

Example

Problem 3.19 showed results for a simple regression of dividends paid per common share in year t (DIVI) as a function of earnings (EARN) per common share in year $(t - 1)$ for AT&T. The problem also included results obtained for a simple regression of dividends (DIVI) as a function of a trend variable (TREND). Suppose now that we are interested in a multiple regression using both earnings and trend as predictor variables. Using the data shown in these two problems, we obtain the following correlation-matrix.

```
> CORR

UPDATING CORR. MATRIX...

                            HOW MANY VARIABLES ?3
```

```
COL. #'S:  ?1,2,3
* # DECIMALS = 3
```

	DIVI	EARN	TREND
DIVI	1.000		
EARN	0.976	1.000	
TREND	0.992	0.965	1.000

Both predictor variables correlate highly with the criterion variable; however, they also correlate highly with each other, as indicated by the correlation coefficient of 0.965. This suggests that it may be difficult to obtain statistically reliable estimates of the separate effects for EARN and TREND in a multiple regression. Indeed, the coefficient for EARN has a low computed t-value, as shown below. However, the multiple regression analysis provides high R^2 and low standard deviation of residuals values.

```
> COEF
```

VARIABLE	B	STD. ERROR(B)	T
EARN	1.6114E−01	1.0604E−01	1.520
TREND	2.5068E−01	5.8826E−02	4.261
CONSTANT	1.8552E+00	3.9366E−01	4.713

```
> SUMM
```

	MULTIPLE R	R−SQUARE
UNADJUSTED	0.9948	0.9896
ADJUSTED	0.9927	0.9854

STD. DEV. OF RESIDUALS =   100583

If we require a t-ratio greater than the computed t-value for EARN, then we may drop EARN from the equation. We are then back to a single regression equation with TREND as the predictor variable, as shown below.

```
> COEF
```

VARIABLE	B	STD.ERROR(B)	T
TREND	3.3690E−01	1.7130E−02	19.667
CONSTANT	2.4414E+00	8.6504E−02	28.223

Note that the estimated standard error for TREND's effect in the multiple regression (0.059, as shown above) is more than three times as large as the corresponding standard error in the simple regression (0.017). Similarly, the standard error for EARN's effect is

0.106 in the multiple regression, compared to only 0.055 in the simple regression (see Chapter 3). The considerable amount of correlation between EARN and TREND is responsible for this increase in uncertainty when both variables are included in a multiple regression.

Extreme Multicollinearity

The effect of correlated predictor variables could have been even more dramatic. Let us continue with the same model but change the data on EARN somewhat. Specifically, let the data for the variables be as follows:

DIVI	EARN 1	TREND
2.80	4.14	1
3.16	4.68	2
3.40	5.22	3
3.70	5.76	4
4.10	6.30	5
4.50	6.84	6
5.00	7.38	7
5.00	7.92	8

The variable EARN has been renamed EARN 1 to reflect the change in the data for this variable. Close examination of the data reveals that the values for EARN 1 increase by exactly 0.54 each year, while TREND increases by exactly 1 each year. Therefore the two variables correlate perfectly. The correlation matrix for all three variables is shown below.

```
> CORR

UPDATING CORR. MATRIX...
                    HOW MANY VARIABLES ?3
COL. #'S: ?1,4,3
* # DECIMALS = 3
                        DIVI          EARN 1         TREND
            DIVI        1.000
            EARN1       0.992         1.000
            TREND       0.992         1.000          1.000
```

We have deliberately created a situation with two perfectly correlated predictor variables. This represents extreme multicollinearity, and it is thus impossible to get a unique solution for this multiple regression (i.e., it is impossible to get separate effects for the two predictor variables). Any computer program will indicate this problem in some manner. The *IDA* program provided an overflow warning, as shown below.

```
> COEF

VARIABLE                 B               STD. ERROR(B)           T

OVERFLOW — WARNING ONLY IN LINE 3580 IN IDA34
   EARN 1
+2.02824E+31
   1.2752E+31      8.2009E+18
+1.55494E+12

OVERFLOW — WARNING ONLY IN LINE 3580 IN IDA34
   TREND
−2.02824E+31
   1.2752E+31      4.4285E+18
−1.55494E+12
```

A Solution to Extreme Multicollinearity

When extreme multicollinearity exists in the data, we have no choice but to (1) remove one of the predictor variables from the model, (2) combine the two predictor variables in some manner, or (3) obtain additional data that may not have such a high degree of correlation between the predictor variables.

Close-to-Extreme Multicollinearity

Changing the data for EARN again, we could also have experienced the following data:

DIVI	EARN 2	TREND
2.80	4.13	1
3.16	4.63	2
3.40	5.17	3
3.70	5.80	4
4.10	6.21	5
4.50	6.83	6
5.00	7.35	7
5.00	7.91	8

EARN 2 describes earnings that increase relatively consistently but not by a constant amount. The values increase by amounts between 0.41 and 0.63. Nevertheless, the correlation between EARN 2 and TREND is very high, as shown below.

```
> CORR

HOW MANY VARIABLES ?3
COL. #'S: ?1,5,3
* # DECIMALS = 3
```

	DIVI	EARN2	TREND
DIVI	1.000		
EARN2	0.991	1.000	
TREND	0.992	0.999	1.000

This situation permits a unique solution for the estimated separate effects of the predictor variables on the criterion variable. But the estimated standard errors tend to be large, and the estimated effects are thus unreliable. The results are shown below.

```
> COEF
```

VARIABLE	B	STD. ERROR(B)	T
EARN2	−.5503E+00	1.1432E+00	−0.481
TREND	6.3492E−01	6.1934E−01	1.025
CONSTANT	4.4053E+00	4.0806E+00	1.080

```
> SUMM
```

	MULTIPLE R	R−SQUARE
UNADJUSTED	0.9927	0.9854
ADJUSTED	0.9897	0.9796

STD. DEV. OF RESIDUALS = .11889

```
> ANOV
```

SOURCE	SS	DF	MS	F
REGRESSION	4.77048E+00	2	2.38524E+00	168.75
RESIDUALS	7.06740E−02	5	1.41348E−02	
TOTAL	4.84115E+00	7	6.91593E−01	

The test for the model's overall statistical significance suggests that the null hypothesis (all β values equal zero) can be rejected, because the computed value for F equals 168.75 ($F_{0.05,2,5} = 5.79$). Nevertheless, *both* predictor variables have low computed t-values, suggesting that neither coefficient is statistically significantly different from zero in the multiple regression. Also, EARN2 has a negative coefficient, a result that is substantively different from what we expect. This occurs because we cannot obtain reliable estimates of the predictor variables' separate effects in this case.

Again, multicollinearity requires a change in the model. The simplest change is to drop one predictor variable. We could also combine the predictor variables or use other procedures for estimating the separate effects.[7] Unfortunately, recognizing multi-collinearity is not always a simple task. For example, it is impossible to say that a correlation of 0.95 between two predictor variables indicates a multicollinearity problem. Whether a problem exists depends on several other factors as well. The problem caused by a correlation of 0.95 between the predictor variables is greater for a small sample size than for a large sample size. This is consistent with the notion that getting more observations may help reduce the problem of multicollinearity. More precisely, as shown in Table 4.2, the effect of an increase in the correlation between two predictor variables can be offset, by having more variation in the predictor variable(s).

We can see the effect of a high correlation between two predictor variables on the standard error of a slope coefficient more clearly with a transformation of the formula for the standard error. We start with the formula for s_{b1} shown earlier:

$$s_{b_1} = \frac{s}{\sqrt{\dfrac{\Sigma x_{1i}^2 \Sigma x_{2i}^2 - (\Sigma x_{1i} x_{2i})^2}{\Sigma x_{2i}^2}}}$$

$$= \frac{s}{\sqrt{\Sigma x_{1i}^2 - \dfrac{(\Sigma x_{1i} x_{2i})^2}{\Sigma x_{2i}^2}}}$$

$$= \frac{s}{\sqrt{\Sigma x_{1i}^2 = \dfrac{r_{X_{1i}X_{2i}}^2 \Sigma x_{1i}^2 \Sigma x_{2i}^2}{\Sigma x_{2i}^2}}} \quad \text{since} \quad r_{X_{1i}X_{2i}}^2 = \frac{(\Sigma x_{1i} x_{2i})^2}{\Sigma x_{1i}^2 \Sigma x_{2i}^2}$$

$$= \frac{s}{\sqrt{\Sigma x_{1i}^2 (1 - r_{X_{1i}X_{2i}}^2)}}$$

where $r^2_{X_{1i}X_{2i}}$ is the squared correlation coefficient between X_{1i} and X_{2i}.

Thus, we can express the estimated standard error for one slope coefficient as a function of the squared correlation between the predictor variables. The higher this correlation, the higher the standard error. We show this relation in Table 4.2 for selected values of the (squared) correlation between the predictor variables.

If s remains approximately constant, the entries in the last column show how much the variation for the first predictor variable in the sample has to increase to compensate for the higher correlation between the two predictor variables. For example, a correlation of 0.50 has little impact on the standard error. But a correlation of 0.95 requires about three times as much variation in X_1 (or three times the sample size[8]) compared to having zero correlation. With a correlation of 0.99 we require about seven times as much variation, or seven times the sample size. And with a correlation of 0.999

Table 4.2
The effect of increasing correlation between two predictor variables on the standard error of the slope coefficient

$r^2_{x_{1i}x_{2i}}$	$r_{x_{1i}x_{2i}}$	s_{b_1}
0	0	$\dfrac{s}{\sqrt{\Sigma x^2_{1i}}}$
0.25	0.50	$1.2\,\dfrac{s}{\sqrt{\Sigma x^2_{1i}}}$
0.64	0.80	$1.7\,\dfrac{s}{\sqrt{\Sigma x^2_{1i}}}$
0.90	0.95	$3.2\,\dfrac{s}{\sqrt{\Sigma x^2_{1i}}}$
0.98	0.99	$7.1\,\dfrac{s}{\sqrt{\Sigma x^2_{1i}}}$
0.998	0.999	$22.4\,\dfrac{s}{\sqrt{\Sigma x^2_{1i}}}$

we need about twenty-two times the sample size. (Or if a sample size of ten is acceptable in case of zero correlation between two predictors, we need a sample size of 224 to get the same precision for the slope coefficients.)

To understand the seriousness of multicollinearity for a given set of data, we recommend comparing a model with both predictor variables to a model with one predictor variable. This comparison allows us to see the effect on the estimated standard error for one or the other predictor variable. Multicollinearity is also indicated when statistical significance is achieved for the model as a whole (based on an F-test) without obtaining statistical significance for any of the separate effects. This was the case in the model with EARN 2 as a predictor variable. Multicollinearity is discussed further in Chapter 5, when more than two predictor variables are included in the model.

Problems

4.13 Use the formula for s_{b_1}, used in Table 4.2, to verify the effect of collinearity on the standard error for TREND in the original multiple regression $\text{DIVI} = a + b_1\,\text{TREND} + b_2\,\text{EARN}$. Note, however, that the standard error of residuals differs slightly between the simple regression analysis when TREND is the only predictor variable ($s = 0.111$, see Chapter 3) and a multiple regression analysis ($s = 0.101$).

4.14 Use the formula for s_{b_1}, used in Table 4.2, to verify the effect of collinearity on the standard error for TREND in the original multiple regression $\hat{DIVI} = a + b_1$ TREND $+ b_2$ EARN2. Note, however, that the standard error of residuals differs slightly between the simple regression analysis when TREND is the only predictor variable ($s = 0.111$, see Chapter 3) and a multiple regression analysis ($s = 0.119$).

Specification Error

Omitting a relevant predictor variable can cause severe problems, as we have argued. Such an omission is an example of what is often called a specification error. In general, if the model specified is an incorrect representation of reality, then the results are subject to the consequences of specification error. In this section we formalize this notion and show the consequences for the estimate of a slope parameter when one predictor variable is omitted. We also show the consequences of including an irrelevant predictor variable in the model.

Suppose the true or correct model is

$$Y_i = \alpha + \beta_1 X_{1i} + \beta_2 X_{2i} + u_i$$

but instead, we actually use

$$Y_i = \alpha + \beta_1 X_{1i} + v_i$$

where v_i is an error term (that differs from u_i).

From the simple regression, based on Y and X_1, we obtain

$$b_1 = \frac{\sum X_{1i} Y_i - \dfrac{\sum X_{1i} \sum Y_i}{n}}{\sum X_{1i}^2 - \dfrac{(\sum X_{1i})^2}{n}}$$

If the correct model includes X_2, as shown above, then

$$E(b_1) = \beta_1 + \beta_2 \frac{\sum X_{1i} X_{2i} - \dfrac{\sum X_{1i} \sum X_{2i}}{n}}{\sum X_{1i}^2 - \dfrac{(\sum X_{1i})^2}{n}}$$

$$= \beta_1 + \beta_2 c_1$$

where
$$c_1 = \frac{\Sigma X_{1i} X_{2i} - \dfrac{\Sigma X_{1i} \Sigma X_{2i}}{n}}{\Sigma X_{1i}^2 - \dfrac{(\Sigma X_{1i})^2}{n}}$$

Note that c_1 is the slope coefficient for a simple regression analysis of X_2 as a function of X_1. That is, c_1 is the slope coefficient in the equation

$$\hat{X}_{2i} = c_o + c_1 X_{1i}$$

If, as we assume, X_2 matters, then $\beta_2 \neq 0$. And $c_1 \neq 0$, unless the sample correlation between X_1 and X_2 is precisely zero. Thus, if it is incorrect to omit X_2 from the model, the estimator b_1 (from the simple regression) is biased. The bias is the difference between $E(b_1)$ and β_1, or $\beta_2 c_1$. In general, the greater the impact of X_2 on Y, holding X_1 constant, and the greater the correlation between X_1 and X_2, the greater the bias.

Now suppose that we include an irrelevant variable in a model. Specifically, let

$$Y_i = \alpha + \beta_1 X_{1i} + u_i$$

be the correct model, and suppose we actually use

$$Y_i = \alpha + \beta_1 X_{1i} + \beta_2 X_{2i} + v_i$$

Based on the multiple regression, with the variables in deviation form we compute

$$b_1 = \frac{(\Sigma y_i x_{1i})(\Sigma x_{2i}^2) - (\Sigma y_i x_{2i})(\Sigma x_{1i} x_{2i})}{(\Sigma x_{1i}^2)(\Sigma x_{2i}^2) - (\Sigma x_{1i} x_{2i})^2}$$

Now, if the correct model does not involve X_2, the expected value of the slope for X_1 equals the parameter of interest, that is, $E(b_1) = \beta_1$. Thus, the estimator for β_1 is still unbiased, even if there are superfluous predictor variables in the model. However, the multiple regression does not give the minimum variance estimator; rather, the estimator from the simple regression is the minimum variance estimator. Nevertheless, it is better to err on the side of including variables which might be irrelevant than to err by omitting variables which belong in the model. Of course, the expected value of b_2 is zero if X_2 is irrelevant. Hence, the t-ratio for the slope coefficient of X_2 should be small. If X_2 is indeed irrelevant, we expect to exclude it from the equation because of its insignificant t-ratio.

SUMMARY

The problem of omitting a relevant predictor variable from the model is similar to not using the best sampling procedure, as we have suggested in Table 4.1. To use the best sampling procedure for a given problem, we need expertise about sampling theory as well as about the substantive problem. Similarly, to make proper use of regression analysis, we need expertise in the theory of regression and the substantive area of application. The theory helps us understand the consequences of problems such as the omission of a relevant predictor variable from the model. But knowledge about the substantive area is *required* to know whether we may have omitted a relevant predictor variable from the model.

In this chapter we have shown a multiple regression analysis with two predictor variables. We first demonstrated, graphically and by example, how a second predictor variable can be held constant in order to estimate the effect of the first predictor variable. We then showed the formula for the slope coefficient for a multiple regression analysis. This formula is substantially more complex than the slope coefficient formula in a single regression analysis. The added complexity results from the requirement that the effect of one predictor variable is estimated, while the other predictor variable is held constant.

We have also argued that the need for a multiple regression analysis (as opposed to simple regression) increases as the correlation between the predictor variable of primary interest and another relevant predictor variable increases. Ironically, however, it is also true that the greater this correlation, the more difficult it is to obtain a reliable result from the multiple regression analysis. Thus, the *validity* (lack of bias) requires the inclusion of all relevant predictor variables, while the *reliability* (small standard error) of the parameter estimates may worsen if the predictor variables are highly correlated.

We cannot emphasize too strongly the importance of substantive knowledge for proper model specification (e.g., identification of relevant predictor variables). Thus, regression analysis is primarily useful for the estimation of the magnitude of effects, not for the determination of which predictor variables are relevant. Only to a limited extent can we use the residuals from a model as an aid in our efforts to identify problems with the model specification. Unfortunately, the value of R^2 (unadjusted or adjusted) is not helpful for determining whether relevant predictor variables are omitted. Similarly, whether estimates have the expected sign does not guarantee their validity. (The slope coefficient for price, for example, should be negative.) We discuss such considerations relevant to establishing model validity in more detail in subsequent chapters.

APPENDIX 4.1
DERIVATIVES TO SOLVE FOR a, b_1, AND b_2

Multiple regression solution for an equation with two predictor variables

$$Y_i = \alpha + \beta_1 X_{1i} + \beta_2 X_{2i} + u_i$$

$$\hat{u}_i = Y_i - \hat{Y}_i$$
$$= Y_i - a - b_1 X_{1i} - b_2 X_{2i}$$

$$\hat{u}_i^2 = (Y_i - a - b_1 X_{1i} - b_2 X_{2i})^2$$

$$\Sigma \hat{u}_i^2 = \Sigma (Y_i - a - b_1 X_{1i} - b_2 X_{2i})^2$$

$$\frac{\partial \Sigma \hat{u}_i^2}{\partial a} = -\Sigma(Y_i - a - b_1 X_{1i} - b_2 X_{2i}) = 0$$

$$\frac{\partial \Sigma \hat{u}_i^2}{\partial b_1} = -\Sigma X_{1i}(Y_i - a - b_1 X_{1i} - b_2 X_{2i}) = 0$$

$$\frac{\partial \Sigma \hat{u}_i^2}{\partial b_2} = -\Sigma X_{2i}(Y_i - a - b_1 X_{1i} - b_2 X_{2i}) = 0$$

Simplifying these three derivatives we obtain

$$\Sigma Y_i = na + b_1 \Sigma X_{1i} + b_2 \Sigma X_{2i}$$

$$\Sigma X_{1i} Y_i = a\Sigma X_{1i} + b_1 \Sigma X_{1i}^2 + b_2 \Sigma X_{1i} X_{2i}$$

$$\Sigma X_{2i} Y_i = a\Sigma X_{2i} + b_1 \Sigma X_{1i} X_{2i} + b_2 \Sigma X_{2i}^2$$

From these three equations we can find solutions for a, b_1, and b_2. In *deviation* form (i.e., $y_i = Y_i - \bar{Y}$; $x_{1i} = X_{1i} - \bar{X}_1$; $x_{2i} = X_{2i} - \bar{X}_1$), the solution for b_1 and b_2 is

$$b_1 = \frac{(\Sigma y_i x_{1i})(\Sigma x_{2i}^2) - (\Sigma y_i x_{2i})(\Sigma x_{1i} x_{2i})}{(\Sigma x_{1i}^2)(\Sigma x_{2i}^2) - (\Sigma x_{1i} x_{2i})^2}$$

$$b_1 = \frac{(\Sigma y_i x_{2i})(\Sigma x_{1i}^2) - (\Sigma y_i x_{1i})(\Sigma x_{1i} x_{2i})}{(\Sigma x_{1i}^2)(\Sigma x_{2i}^2) - (\Sigma x_{1i} x_{2i})^2}$$

The intercept, using the variables in the original form, is

$$a = \bar{Y} - b_1 \bar{X}_1 - b_2 \bar{X}_2$$

The formula for b_1 estimates the effect of X_1 on Y, holding X_2 constant. Similarly, the formula for b_2 estimates the effect of X_2 on Y, holding X_1 constant. Finally, a is the intercept, the estimated value for the criterion variable when both X_1 and X_2 are zero.

The formula for b_1 shown above, in comparison to the one for an equation with only one predictor variable, is considerably more involved; hence, computers are normally used to estimate the unknown values in a model with multiple predictor variables.

APPENDIX 4.2
EXPRESSIONS FOR UNADJUSTED R^2

For a multiple regression with two predictor variables, the unadjusted R^2 for the estimated equation can be calculated as follows:

$$R^2 = \frac{r_{Y,X_1}^2 + r_{Y,X_2}^2 - 2r_{Y,X_1}\,r_{Y,X_2}\,r_{X_1,X_2}}{1 - r_{X_1,X_2}^2}$$

If the variables X_1 and X_2 are uncorrelated (i.e., $r_{X_1,X_2} = 0$) then the formula reduces to

$$R^2 = r_{Y,X_1}^2 + r_{Y,X_2}^2$$

APPENDIX 4.3
ESTIMATED STANDARD ERROR OF PREDICTION

Under the error-term assumptions, and based on the model

$$\hat{Y}_0 = a + b_1 X_{10} + b_2 X_{20}$$

the estimated standard error for an individual prediction is

$$s_{\hat{Y}_o} = s\sqrt{1 + \frac{A}{B}}$$

where

$$s = \sqrt{\frac{\Sigma(Y_i - \hat{Y}_i)^2}{n - 3}}$$

$$A = \Sigma X_{1i}^2 \Sigma X_{2i}^2 - (\Sigma X_{1i}\Sigma X_{2i})^2 + 2X_{10}\Sigma X_{1i}[(\Sigma X_{2i})^2 - \Sigma X_{2i}^2]$$
$$+ 2X_{20}\Sigma X_{2i}[(\Sigma X_{1i})^2 - \Sigma X_{1i}^2] + 2X_{10}X_{20}[\Sigma X_{1i}\Sigma X_{2i}(1 - n)]$$
$$+ X_{10}^2[n\Sigma X_{2i}^2 - (\Sigma X_{2i})^2] + X_{20}^2[n\Sigma X_{1i}^2 - (\Sigma X_{1i})^2]$$

$$B = n\Sigma X_{1i}^2 \Sigma X_{2i}^2 + (2 - n)(\Sigma X_{1i}\Sigma X_{2i})^2 - (\Sigma X_{1i})^2\Sigma X_{2i}^2 - (\Sigma X_{2i})^2\Sigma X_{1i}^2$$

The estimated standard error for a conditional mean prediction is

$$s_{\hat{\bar{Y}}_o} = s\sqrt{\frac{A}{B}}$$

where s, A, and B are defined as shown above.

ENDNOTES

1. If the null hypothesis can be rejected for $\beta_1 = 0$, then it can be rejected for all values of $\beta_1 > 0$, given the use of the one-tailed test.

2. *Interactive Data Analysis*, a computer program written at the Graduate School of Business, University of Chicago.

3. Note that the formulas for the standard error of predictions in Chapter 3 apply only to a *simple* linear regression model. To compute the standard error of prediction in a multiple regression, a more involved formula is needed (see Appendix 4.3).

4. All computer output shown here was obtained from *IDA*.

5. *IDA* also provides standardized regression coefficients (beta weights). We have eliminated that part of the output. See also Chapter 8.

6. A reliable estimate is an estimate with a small standard error. A valid estimate is an estimate that is unbiased.

7. See Marquardt, Donald W., and Ronald D. Snee, "Ridge Regression in Practice," *The American Statistician*, 29 (February 1975), 3–19.

8. Expressing the problem in terms of the larger sample size required assumes, of course, that the amount of variation in the predictor variable increases proportionally with increases in the sample size.

REFERENCE

Marquardt, Donald W., and Ronald D. Snee. "Ridge Regression in Practice." *The American Statistician* 29 (February 1975): 3–19.

5

Multiple Regression Analysis with *m* Predictor Variables

To generalize the regression model for problems involving any number of predictor variables, we use

$$Y_i = \alpha + \beta_1 X_{1i} + \beta_2 X_{2i} + \cdots + \beta_m X_{mi} + u_i$$

This model is often referred to as the *general linear model*. It is *general* because it allows for an arbitrary number, *m*, of predictor variables. And, for each of the *m* predictor variables specified, the effects are assumed to be *linear*. However, it is possible to allow for nonlinear effects by transforming one or more variables. In Chapter 6 we discuss how the general linear model can be used to allow for nonlinear effects. With such transformations the model allows for linear effects on the transformed variables but nonlinear effects on the original variables.

Apart from the linearity implicit in this model, there are no interaction effects accommodated in this model (unless, again, there are transformations or other manipulations that allow for interactions). Thus, the effect of a given predictor variable, say X_1, is always β_1, regardless of the level of X_2 or any other predictor variable. In this chapter we discuss how interaction effects can be accommodated specifically in the context of indicator variables.

For the general linear model the estimates a, b_1, b_2, ..., b_m are obtained with complex and involved formulas (see Appendix 5.2 for a treatment, using matrix algebra). As before, we desire to choose estimates so that the sum of squared residuals is minimized. The result is a system of $(m + 1)$ equations or formulas that can be used to compute the desired coefficients based on a set of sample data. For unique solutions to exist, we require that the equations are *linearly independent*. If this condition is violated, a multiple regression computer program will abort, as happened in the case of extreme multicollinearity between two predictor variables in Chapter 4.

Extreme Multicollinearity

No solution to a multiple regression problem can be obtained if the system of $(m + 1)$ equations is linearly dependent. This is equivalent to saying that there is extreme multicollinearity between the predictor variables. For example, if any two predictor variables are perfectly correlated, there is no unique solution for the multiple regression analysis. For that reason, many regression users first inspect the simple correlation matrix to see if there are any correlation coefficients equal to $+1$ or -1 between the predictor variables. Unfortunately, however, the lack of a perfect correlation between all pairs of predictor variables does not guarantee the existence of a unique solution.

To see why the problem is more general than is indicated by the simple correlation coefficients, we use the following intuitive argument. If we were able to see the formula for a given slope coefficient, say b_1, we would see that the formula involves the covariation or correlation between that predictor variable and all other predictor variables in the model. This formula provides an estimate of the effect of X_1 on Y, while holding all other predictor variables *simultaneously* constant. To do this, X_1 cannot be perfectly correlated with any linear combination of all other predictor variables. The simple correlation coefficient between, say, X_1 and X_2 gives, therefore, an incomplete picture, because it is the correlation between X_1 and a special linear combination of the other predictors (with a weight of one for X_2 and zero for all other variables).

In general, the condition of linear independence for the system of $(m + 1)$ equations is satisfied if none of the predictor variables can be expressed perfectly as a linear combination of the other predictor variables. For example, to determine whether X_1 is perfectly correlated with any linear combination of the remaining predictor variables, we could perform a multiple regression with X_1 as the criterion variable and $X_2, X_3, \ldots, X_m$ as the predictor variables. If this regression analysis results in an unadjusted R^2 value equal to 1.00, it is impossible to obtain a unique solution for the multiple regression analysis of Y as a function of these predictor variables. Thus, if $X_1 = f(X_2, X_3, \ldots, X_m)$ has an R^2 value equal to 1, then there is no unique solution for the slope coefficients and intercept of the model:

$$\hat{Y}_i = a + b_1 X_{1i} + b_2 X_{2i} + \cdots + b_m X_{mi}$$

The same argument applies to a finding of $R^2 = 1$ when any of the other predictor variables are explained as a function of the remaining predictors.

Less-than-Extreme Multicollinearity

In practice, extreme multicollinearity is unlikely to exist. For that reason, we submit the full model for a multiple regression analysis. If there is no unique solution because of extreme multicollinearity, the computer analysis will provide that information. Of course, it is quite possible that a given predictor variable is highly (but not perfectly) correlated with a linear combination of the other predictors. In that case, a unique solution exists, but it tends to be unreliable (i.e., high estimated standard errors for one or more slope coefficients). We want to address two issues here: (1) how to identify the existence of a high degree of multicollinearity, and (2) what to do about it.

Suppose that X_1 is highly correlated with a linear combination of the other predictor variables in the model. As a result, the estimated standard error of b_1 is fairly high, because it is influenced by this correlation. (See Table 4.2, for a case with only two predictor variables.) Specifically, the *t*-ratio for b_1 may be so small that it is impossible to reject the null hypothesis that $\beta_1 = 0$ at an acceptable level of significance. Yet our substantive understanding of the problem we are attacking may lead us to have high expectations about the relevance of X_1. Should we eliminate X_1 from the equation, because of the low *t*-ratio?

In terms of identifying the existence of a high degree of multicollinearity, we can use the regression with X_1 as the criterion variable and $X_2, \ldots, X_m$ as the predictor variables. However, it is also possible to do the following. First, estimate the complete model with Y as a function of all m predictor variables. Next, eliminate one of the predictor variables (for example, the one with the highest absolute simple correlation with X_1), and re-estimate the model. Then, in the re-estimated model, the estimated standard error of b_1 should be considerably reduced, if there is a high degree of multicollinearity in the complete model. The greater this reduction, the greater the correlation between X_1 and the predictor variable eliminated from the model. Thus, in general a comparison of the estimated standard errors for a given slope coefficient between two multiple regression analyses (one containing all m predictors, the other containing $(m - 1)$ predictors) provides information about the extent of multicollinearity in the full model due to the eliminated predictor variable. We now discuss what to do about this collinearity, with suggested remedies dependent on the objectives of the study.

Forecasting

If the purpose of the study is to develop a forecasting model, multicollinearity is often not an issue. For example, suppose X_1 is eliminated from the model because it is highly correlated with a linear combination of the other predictor variables. Now, if X_1 is truly a relevant predictor, the parameter estimates for the remaining predictors tend to be biased. Nevertheless, the forecasts produced by this otherwise deficient model may still be accurate. In particular, most of the explanatory power of X_1 is contained in one or more other predictor variables. As long as this correlation between X_1 and the other predictors continues to exist, the forecasts from this deficient model will not be systematically affected. On the other hand, if this correlation changes, the forecasts will be affected by the model's deficiency. (This happens especially when the predictor variables are controlled by a manager or a policy maker.) The reason for this is that the elimination of a relevant predictor variable biases the coefficients of the remaining predictor variables. The nature and magnitude of this bias depends, among other things, on the degree of multicollinearity. If this degree of multicollinearity changes, then the nature of the bias changes, thereby affecting the forecasting accuracy.

Understanding

If the objective is to understand the relationship between a criterion variable and several predictor variables, it is critical that all relevant variables are included in the model. The presence of multicollinearity may make it difficult or impossible to obtain the desired degree of reliability (i.e., low standard errors) for the coefficients. As a result the computed t-ratio for one or more predictor variables may be below the critical value (based on an appropriate type 1 error probability).

We have seen in Chapter 4 that multicollinearity may cause all t-ratios to be small, even when the model as a whole is statistically significant. Thus, the elimination of one predictor variable (perhaps the one with the smallest absolute t-ratio) can dramatically change the t-ratios for the remaining predictors. We recommend, therefore, that after the full model is estimated, only one predictor is eliminated at a time. The (reduced) model can then be estimated, and the t-ratios for this model can form the basis for any further elimination of predictor variables.

Solutions to Multicollinearity

We discuss briefly several approaches available for the resolution of severe multicollinearity:

1. Exclude a predictor variable with a low t-ratio
2. Obtain more data relevant to the problem
3. Reformulate the model with the specific objective to decrease multicollinearity
4. Use a procedure specifically developed for cases with severe multicollinearity

The first possible solution suggests that a predictor variable with a statistically insignificant t-ratio be eliminated. In general, an insignificant t-ratio indicates that the predictor variable is irrelevant or that it has an effect, but we are unable to obtain a reliable estimate. In either case, it seems appropriate to eliminate such a predictor variable from the model.

A better solution, of course, is to obtain more data. With more data, especially data with a smaller degree of multicollinearity, it is more likely that a significant effect can be obtained if the predictor variable that is otherwise eliminated from the model is truly relevant. Of course, the opportunity to add more data to the sample is often quite limited.

A third possible solution is to reformulate the model. In some sense, the elimination of one predictor variable amounts to a model reformation. But we may also combine two predictor variables and in that manner resolve the problem. However, this is appropriate only if the two variables are more or less substitute measures of the same underlying construct. For example, suppose that the observations for a model represent individuals, and that two of the predictors measure age and work experience. These two variables tend to be correlated, and may both measure individuals' learned skills for a certain job as well as their maturity. It may be sufficient to use one of these variables or to define a new predictor variable that combines the two predictor variables.

If the source of multicollinearity is two or more predictor variables that measure different aspects, such combinations are not appropriate. However, it may be possible to redefine the variables. For example, if the sample represents time-series data, the model in period t is

$$Y_t = \alpha + \beta_1 X_{1t} + \beta_2 X_{2t} + \cdots + \beta_m X_{mt} + u_t$$

Similarly, in period $(t - 1)$ we have

$$Y_{t-1} = \alpha + \beta_1 X_{1,t-1} + \beta_2 X_{2,t-1} + \cdots + \beta_m X_{m,t-1} + u_{t-1}$$

And by subtracting one equation from the other, we can state the model in terms of changes over time:

$$(Y_t - Y_{t-1}) = \beta_1(X_{1t} - X_{1,t-1}) + \beta_2(X_{2t} - X_{2,t-1})$$
$$+ \cdots + \beta_m(X_{mt} - X_{m,t-1}) + u_t - u_{t-1}$$

Note that for the reformulated model, there is no intercept. Apart from that, the model contains the same parameters. Importantly, this reformulation may not have the same degree of multicollinearity as the original model. That is, the multicollinearity in a set of variables may change considerably by expressing the same variables in terms of first differences. Such opportunities for model reformulation are useful to consider when we encounter severe multicollinearity.

In case none of the possible solutions suggested above are feasible or appropriate, we may consider using an alternative estimation procedure. Probably the most popular procedure is *ridge regression analysis*.[1] Essentially, this technique provides biased estimates with substantially reduced standard errors, relative to least squares. The idea is that it may be better to allow for a small amount of bias in order to have a substantially reduced amount of uncertainty (low standard errors for the slope coefficients).

A Test of the Equation as a Whole

As we discussed in Chapter 4, the first statistical test to employ is the test of the equation as a whole. Especially when we have many predictor variables in the model, there is a high probability that one or more slope coefficients have t-ratios above the critical value, even if the model as a whole has no significant explanatory power. Compare, for example, the practice of determining the percentage of times each electoral district in the

United States has voted with the winner of a presidential electron. Does the district with the highest percentage provide superior predictive power? Certainly the observed percentage is biased upward due to the fact that by chance one particular district is likely to do substantially better than, say, fifty percent. Before we believe in the superior predictive power of this district, we must determine whether the observed percentages are significantly different from what could be expected, given the number of elections and districts, if all districts are equally capable of voting with the winner. For example, if we toss each of ten coins twenty times, the percentage of heads will also differ substantially between the coins. Yet, we usually do not believe that the coins differ systematically in the likelihood of a head resulting. In other applications, a high degree of multicollinearity may prevent the t-ratios from being above the critical value, while the model is statistically significant. An example of this is shown in Chapter 4, for a case involving very high correlation between two predictor variables.

Under the null hypothesis of no effect for all of the predictor variables

$$H_o: \beta_1 = \beta_2 = \cdots = \beta_m = 0$$

and under the four assumptions about the error term (see Chapter 3), the ratio

$$\frac{R^2/m}{(1 - R^2)/(n - m - 1)}$$

is distributed according to the F-distribution with m and $(n - m - 1)$ degrees of freedom. If the computed F-value exceeds the tabulated value, we conclude that the obtained result is unlikely to occur if the null hypothesis is true. Based on this outcome, we reject the null hypothesis in favor of the alternative hypothesis

H_A: at least one of the slope parameters is different from zero.

Conceptually the formula for the F-test captures the proportion of the variation explained by the estimated equation divided by the proportion of the variation which is unexplained. And each proportion is adjusted (divided) by the appropriate degrees of freedom. The explained variation is adjusted by the number of predictor variables used, while the unexplained variation is adjusted by the degrees of freedom left or unused. If the explanatory power of the predictor variables is truly zero, the ratio of explained over unexplained is expected to be approximately 1.00. Thus, the higher this ratio, the more reason we have to believe that the model has explanatory power beyond the sample data.

We have emphasized that the F-test is used before decisions are made about the statistical significance of individual slope coefficients. However, this does not address to which model the F-test should be applied. For example, in an application we may specify a model with many predictor variables. However, the final model selected might contain only a subset of these variables. That is, we may have eliminated some predictor variables, based on the values for the t-ratios.

The *F*-test for the equation should be applied to the first model estimated. Usually, this is the full model, containing all predictor variables expected to be relevant. If this initial model explains a significant amount of variation in the criterion variable, it is then appropriate to focus separately on the individual tests (*t*-ratios) for each of the predictor variables. If the *F*-test is not significant, we cannot have confidence in the initial model nor in a simplified model (using a subset of the predictor variables).

In empirical work, it is often unavoidable to use iterative procedures for the purpose of identifying an appropriate reduced model. Such ad hoc analysis tends to result in an overstatement of the explanatory and predictive power of the final model. For that reason, it is important to use a validation or holdout sample (see Chapter 10), whenever iterations are used to identify the "best" model.

Sample Size

The *F*-test formula makes it apparent that the computation is meaningful only if $(n - m - 1)$ is a positive number. In the simplest case, when $m = 1$, this means that we need $n \geq 3$. Intuitively it is easy to see why this is so. With only two observations, we can fit a straight line perfectly through the data points. But if we believe that the true relationship is not perfect, we have no opportunity to assess this uncertainty. Similarly, if we want to use two predictor variables, we can obtain a perfect fit ($R^2 = 1$) based on only three observations. Of course, we usually desire the number of observations to be substantially greater than the number of predictor variables. Some researchers suggest that the number of observations should be three or four times as large as the number of predictor variables. However, it is difficult to justify a particular rule of thumb for this question. For example, as the multicollinearity in the sample data increases, more information is required to obtain the desired amount of precision for the individual slope coefficients.

Tests about Slope Parameters

After the *F*-test is conducted on the initial model and the null hypothesis that all slope parameters are zero is rejected, individual terms of the model can be evaluated. Assuming that the four assumptions about the error term are valid, we can use the computed *t*-ratio for an individual slope coefficient for a more specific statistical test. For the *j*th slope parameter, β_j, the ratio

$$\frac{b_j - \beta_j}{s_{b_j}}$$

follows the *t*-distribution with $(n - m - 1)$ degrees of freedom. For the null hypothesis

H_o: $\beta_j = 0$ this ratio can be simplified to

$$\frac{b_j}{s_{b_j}}$$

Thus, the computed t-ratio can be used as a basis for determining whether there is evidence of a (linear) effect of X_j, the jth predictor variable, on the criterion variable, holding the other predictors constant. If the computed t-value is below the critical value, we may delete this predictor variable from the model.

We note, however, that there are many reasons why the t-ratio for a given slope coefficient may be insignificant. First, the predictor variable has an effect that is different from the functional form assumed (incorrect functional form). Second, the model excludes other relevant predictor variables (omitted variables). Third, the predictor variable is highly correlated with one or more other predictor variables included in the model (multicollinearity). Fourth, the predictor variable has no relation with the criterion variable (irrelevance).

Insignificance due to either of the first two reasons requires us to investigate the existence of superior functional forms and additional predictor variables. The third reason requires the exploration of additional data sources, model reformulation, or alternative estimation procedures. Only the fourth reason is proper justification for eliminating a predictor variable from the model. We have to be careful to consider each of these reasons before eliminating a predictor variable.

Confidence Intervals for Predictions

Once a model has been tested, it is appropriate to consider it for predictions. In case of a conditional mean, we use

$$\hat{\hat{Y}}_o = a + b_1 X_{10} + b_2 X_{20} + \cdots + b_m X_{mo}$$

where   the subscript o indicates that the prediction is based on a set of prespecified values for the predictor variables.

The uncertainty about this prediction can be quantified as follows:

$$\hat{\hat{Y}}_o \pm (t_{n-m-1}) s_{\hat{\hat{Y}}_o}$$

where t_{n-m-1} is the tabulated t-value, given a specified confidence level and $(n - m - 1)$ degrees of freedom.

Similarly, for individual predictions, we have

$$\hat{Y}_o = a + b_1 X_{10} + b_2 X_{20} + \cdots + b_m X_{mo} \quad \text{and}$$

$$\hat{Y}_o \pm (t_{n-m-1}) s_{\hat{Y}_o}$$

We note that the formulas discussed in Chapter 4 for s_{b_1} as well as for $s_{\hat{\bar{Y}}}$ and $s_{\hat{Y}_o}$ are not applicable here. Computer software is available that provides the desired values directly.

Indicator Variables

In empirical work, we frequently have categorical predictor variables. For example, suppose that in a study of the relation between macroeconomic variables, military activity is considered a relevant predictor variable. Although it may be possible to measure military activity by expenditures (a continuous variable), we often assume that the effect is appropriately captured by a categorical variable that distinguishes between periods of war and peace. In regression analysis, such a variable is called an indicator or dummy variable. We provide a detailed treatment of indicator variables below, using starting salary for graduates of an academic program as an example.

Two Categories

Suppose we believe that gender affects starting salary. In fact, let us suppose that it is the only relevant predictor variable, and we are interested in determining whether on the average there is a difference in starting salaries between females and males. One way to investigate this would be to compute the two sample means, female and male, and to use a *t*-test for the difference between them. We consider the same issue in the context of regression analysis, with one indicator variable:

$$X = 1 \text{ if female}$$
$$= 0 \text{ if male}$$

The model is

$$Y_i = \alpha + \beta X_i + u_i$$

where Y_i is starting salary for the *i*th student

The estimated equation is

$$\hat{Y}_i = a + b X_i$$

Suppose that we want to predict a male student's starting salary. In that case $X = 0$, and $\hat{Y}_M = a$. Thus, the intercept is the sample mean for male (M) students. For a female student $X = 1$, and $\hat{Y}_F = a + b$. The sample mean for female (F) students is therefore indicated by the sum of the intercept and the slope coefficient.[2] To test for the difference between the sample means, we require $(\hat{Y}_F - \hat{Y}_M) = (a + b) - a = b$. Hence, the slope coefficient, b, is the estimated difference between the female and male mean starting salaries. To test a null hypothesis of no difference we define

H_O: $\beta = 0$ (The difference between the two population means is zero.)

H_A: $\beta \neq 0$ (The difference between the two population means is not zero.)

If the error-term assumptions are satisfied,

$$\frac{b - \beta}{s_b}$$

follows a t-distribution with $(n - 2)$ degrees of freedom. Thus the t-ratio for the indicator variable's coefficient can be used to test the difference between the two means.

We show in Figure 5.1 a plot of (hypothetical) data for males and females. In this Figure, the slope is the difference between the average Y-value for males and females. Since there are only two possible values for X, the scattergram in Figure 5.1 of Y versus

Figure 5.1
Difference between two sample means (where $\hat{Y}_F > \hat{Y}_M$)

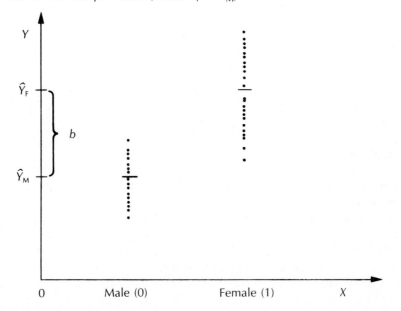

X shows the dispersion of starting salaries about the subgroup sample means separately for males and females.

If we want to determine whether gender is the reason for the observed difference in the two sample means, we have to be concerned with the possibility that the two groups differ in other respects. For example, if starting salary is partly based on academic performance, the difference between the sample means, b, is not a valid indication of the difference created by gender unless the subgroups have equal average academic performance (i.e., if gender is not correlated with academic performance).

We expand the model to include academic performance:

$$Y_i = \alpha + \beta_1 X_{1i} + \beta_2 X_{2i} + u_i$$

where Y_i is starting salary,

X_{1i} is grade point average, and

X_{2i} is an indicator variable equal to 1 if female, 0 if male.

Using multiple regression the estimated model is

$$\hat{Y}_i = a + b_1 X_{1i} + b_2 X_{2i}$$

Separating the subgroups, for a male student we have

$$\hat{Y}_i = a + b_1 X_{1i} \qquad (\text{since } X_{2i} = 0)$$

and for a female student

$$\hat{Y}_i = a + b_1 X_{1i} + b_2 \qquad (\text{since } X_{2i} = 1)$$

Now the difference in average salary between females and males is indicated by the coefficient b_2. This coefficient measures the difference between the two subgroups holding grade-point average (X_1) constant. The model assumes a linear relationship between Y and X_1. The effect of X_2 (gender) is to shift the line of the relationship between salary and grade-point average without changing its slope, as illustrated in Figure 5.2 (assuming that b_2 is positive).

It is important to keep in mind that the value of b_2 is not necessarily the same as the value for b in the simple regression analysis. For a given data set, b measures the difference in average salary between females and males. In the multiple regression analysis, b_2 measures the difference for a given grade-point average. The computed values for b and b_2 will not be the same if males and females differ systematically in their grade-point average values. If males tend to have higher grade-point averages than females, b_2 should be greater than b. On the other hand, if males tend to have lower grade-point averages than females, b_2 should be smaller than b. The value for b_2 depends

Figure 5.2
Difference between two sample means holding X_1 constant

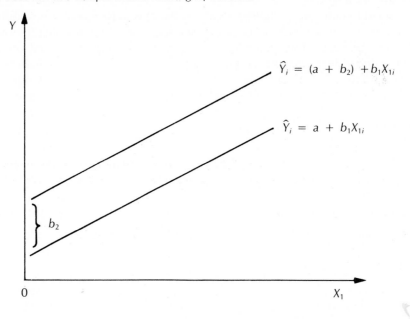

both on the average difference in grade-point average between the two groups and on the correlation between the two predictor variables. Note also that the model for the multiple regression analysis assumes that the difference between females and males is indicated by b_2 regardless of the value for X_1.

Multiple Categories

Now consider a categorical predictor variable with more than two categories. For example, let starting salary be partly determined by the functional classification of the position: accounting, consulting, finance, marketing, and production (assume these are mutually exclusive and collectively exhaustive categories). If we wanted to determine the mean salary for each subgroup in the sample, we could use the following model:

$$Y_i = \alpha + \beta_1 X_{1i} + \beta_2 X_{2i} + \beta_3 X_{3i} + \beta_4 X_{4i} + u_i$$

where the estimated model is

$$\hat{Y}_i = a + b_1 X_{1i} + b_2 X_{2i} + b_3 X_{3i} + b_4 X_{4i}$$

and Y_i = starting salary
       X_{1i} = 1 if the job is in accounting
           = 0 otherwise

$$X_{2i} = 1 \text{ if the job is in consulting}$$
$$= 0 \text{ otherwise}$$
$$X_{3i} = 1 \text{ if the job is in finance}$$
$$= 0 \text{ otherwise}$$
$$X_{4i} = 1 \text{ if the job is in marketing}$$
$$= 0 \text{ otherwise}$$

We emphasize that only four indicator variables are used to distinguish between five categories. Recall that we used only one indicator variable to distinguish between two categories (male, female). In general, we require $(k - 1)$ indicator variables for k categories. The decision as to which category does not have its own indicator variable is arbitrary. However, it is critical to know the exact definition of the indicator variables for proper interpretation of the results.

The reason why there are $(k - 1)$ indicator variables for k categories is the existence of an intercept. That is, to distinguish between five categories, we need to be able to compute five separate means. But with an intercept in the model, we can obtain only four separate expressions of a difference from the intercept value. In our example, production does not have its own indicator variable. As a result the intercept represents this category, and the other categories are evaluated relative to production. This is illustrated below, where we show the relevant parts of the formula for individual predictions of starting salary for a person with the designated classification.

Classification	Prediction
Accounting	$\hat{Y}_i = a + b_1$
Consulting	$\hat{Y}_i = a + b_2$
Finance	$\hat{Y}_i = a + b_3$
Marketing	$\hat{Y}_i = a + b_4$
Production	$\hat{Y}_i = a$

Thus, the slope coefficients measure the predicted difference between someone choosing a job in production and a person choosing a job in another area. The model still contains five parameters (with the intercept) to separate five categories.

Extreme Multicollinearity

Sometimes we mistakenly define more indicator variables than we should. For example, suppose we use five indicator variables for five categories:

$$\hat{Y}_i = a + b_1 X_{1i} + b_2 X_{2i} + b_3 X_{3i} + b_4 X_{4i} + b_5 X_{5i}$$

where $X_{5i} = 1 \text{ if the job is in production}$
$$= 0 \text{ otherwise}$$

and all other variables are as defined before.

Because one indicator variable is redundant, we cannot obtain a unique solution. Our system of six equations (for the six unknowns) is not linearly independent. This is a case of extreme multicollinearity. It happens sometimes in empirical work when we accidentally create too many indicator variables. We show below what the computer file looks like for five individuals, one for each of the categories:

Classification	Data Matrix					
	X_1	X_2	X_3	X_4	X_5	Intercept
Accounting	1	0	0	0	0	1
Consulting	0	1	0	0	0	1
Finance	0	0	1	0	0	1
Marketing	0	0	0	1	0	1
Production	0	0	0	0	1	1

Given that the typical model has an intercept, an additional column is added with all entries equal to one. It is easy to see that this last column is identical to the *sum* of the five indicator variables. Thus, this linear combination of the five indicator variables is perfectly correlated with the column used for the intercept. Only by eliminating one of the indicator variables is it possible to obtain a unique solution. (Alternatively, if we insist on using all five indicator variables, we have to specify the model without an intercept).

A Test of the Equation as a Whole

To determine whether there are statistically significant differences in the average starting salaries between the five categories, we use the F-test. As before, the hypothesis and test statistic are

$$H_o: \beta_1 = \beta_2 = \beta_3 = \beta_4 = 0$$

$$H_A: \text{at least one slope parameter is different from zero}$$

$$F = \frac{R^2/m}{(1 - R^2)/(n - m - 1)}$$

In words, the null hypothesis states that there is no difference in average starting salary between production and any of the other categories. We emphasize, however, that the choice of the base category (production, in this case) has no bearing on the R^2 value or the F-test. If the null hypothesis is rejected, we may proceed to focus on the differences between specific categories.

An Alternative Classification

By using four indicator variables, we also use four degrees of freedom to separate the categories in terms of differences in average starting salaries. It is useful to compare this use of indicator variables with the use of only one variable, as follows:

Classification	X_i
Accounting	1
Consulting	2
Finance	3
Marketing	4
Production	5

In this scheme, we also have a unique representation for each category, while we use only one variable. But if this variable is used as a predictor variable in the equation

$$\hat{Y}_i = a + bX_i$$

then we force a particular estimated difference in the average starting salaries between the groups. Specifically, we have

Classification	Prediction
Accounting	$\hat{Y}_i = a + b$
Consulting	$\hat{Y}_i = a + 2b$
Finance	$\hat{Y}_i = a + 3b$
Marketing	$\hat{Y}_i = a + 4b$
Production	$\hat{Y}_i = a + 5b$

There are two limitations resulting from the use of this variable:

1. If b is positive (negative), production is predicted to have the highest (lowest) salary, marketing the second highest (lowest), and so forth. Instead of allowing the data to indicate *any* differences between the categories, we force the order of the categories (high to low, or low to high) to be production, marketing, finance, consulting, accounting.

2. Even if the particular ordering of the categories implied by this scheme is correct, it is usually not appropriate to force the differences between them to be uniform. The predictions above imply that the differences between all successive categories are indicated by the same b value (i.e., the difference between accounting and consulting equals the difference between consulting and finance, and so forth).

Because of these two limitations, we strongly recommend that this scheme not be used.

Interaction Effects

We have used gender as an indicator variable, along with grade-point average, to explain variation in starting salary. In the example, the indicator variable for gender allowed a shift in the intercept value between males and females. However, the model forced the slope coefficient for grade-point average to be the same for the two categories. If it is not appropriate to force this slope coefficient to be the same, we need to add yet another predictor variable to the model. This third predictor would allow the effect of grade-point average on salary to depend on the gender of the individual. This is commonly referred to as an *interaction* effect.

Actually, we implicitly allow for interaction effects if we carry out two separate regression analyses for the two categories. For example, for the males

$$\hat{Y}_i = a_M + b_M X_i$$

where a_M is the intercept for the males

b_M is the effect of grade-point average on starting salary for the males

Similarly, for the females

$$\hat{Y}_i = a_F + b_F X_i$$

For several reasons, which we discuss below, we may, however, prefer to allow for separate intercepts and slope coefficients in *one* regression analysis. Given that we estimate a total of four parameters in two separate regressions, we must specify a single model with four parameter estimates; for example,

$$\hat{Y}_i = a + b_1 X_{1i} + b_2 X_{2i} + b_3 X_{3i}$$

where $\hat{Y}_i$ is estimated starting salary for the ith individual

X_{1i} is grade-point average for the ith individual

$X_{2i} = 1$ if the ith individual is female

$\quad\quad = 0$ if the ith individual is male

$X_{3i} = X_{1i} * X_{2i}$

This equation reduces to two separate equations for the two categories. For males, $X_{2i} = X_{3i} = 0$, leaving

$$\hat{Y}_i = a + b_1 X_{1i}$$

For females $X_{2i} = 1$ and $X_{3i} = X_{1i}$, which results in

$$\hat{Y}_i = (a + b_2) + (b_1 + b_3)X_{1i}$$

It follows that

$$a = a_M$$

$$(a + b_2) = a_F$$

$$b_1 = b_M$$

$$(b_1 + b_3) = b_F$$

based on a comparison of the two separate regressions, and the separate specifications from the one regression.

Advantages of Using One Equation

Although the estimates of effects are identical for the two approaches described above, there are important differences. For example, the multiple regression equation produces estimates of the differences in intercepts and slope coefficients between the two groups. Specifically, b_2 is the difference in intercept, and b_3 is the difference between the effect of grade-point average for males and females. And the t-ratios associated with these coefficients allow us to determine whether either of these differences is statistically significant.

A single equation has an additional advantage if there are other relevant predictor variables for which we do not allow interaction effects. For example, suppose work experience is added to the equation. For the single equation, we use one more degree of freedom to estimate the slope parameter for experience. But if we use two separate regression analyses for males and females, we lose a total of two degrees of freedom, one for each equation. Clearly if the effect of experience does not depend on another predictor variable (e.g., gender), it is unnecessary and inefficient to estimate this effect separately for each category.

Apart from these advantages, there are other differences. Each separate regression analysis has its own R^2 value and standard deviation of residuals. Of course when one regression analysis (with or without interaction effects) is used, there is only one R^2 value and one standard deviation of residuals that applies to the sample as a whole. Now, it turns out that the standard deviation of residuals for the case of one equation is a weighted average of the standard deviations for the separate regressions. But the R^2 value can be greater in the one multiple regression than both of the separate equations' R^2 values (see Chapter 8). Nevertheless, in terms of true explanatory power, the use of one equation is no better than the two separate equations. Thus, regression analysis offers opportunities to "manipulate" the perceived quality, based on R^2 values, of a model. We return to this problem in Chapter 8.

To summarize, we have three possible models to explain starting salaries:

Model I $\quad \hat{Y}_i = a + b_1 X_{1i}$

Model II $\quad \hat{Y}_i = a + b_1 X_{1i} + b_2 X_{2i}$

Model III $\quad \hat{Y}_i = a + b_1 X_{1i} + b_2 X_{2i} + b_3 X_{3i}$

Note that the addition of a variable may cause the coefficients for other variables to change. Thus, in general, the value for b_1 differs between the three models.

In words, model I allows us to estimate the effect of grade-point average on starting salary and allows no difference between males and females (see Figure 5.3). Model II estimates the same effect, while holding gender constant. That is, the effect of grade-point average is the same for males and females, but the intercept is allowed to be different between the two categories (see Figure 5.4). And model III allows for differences in both intercepts and slope coefficients. The difference in the slope coefficients is an interaction effect, in that the effect of the grade-point average depends on gender (see Figure 5.5).

The models can be compared, based on statistical significance, as follows:

Comparison	H_o	Statistical Significance
II versus I	$\beta_2 = 0$	t-test (based on b_2 in model II)
III versus II	$\beta_3 = 0$	t-test (based on b_3 in model III)
III versus I	$\beta_2 = \beta_3 = 0$	F-test

Figure 5.3
Model I—No gender effect

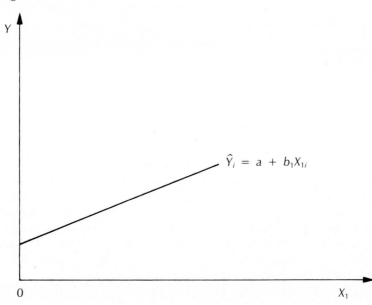

Figure 5.4
Model II—A main effect for gender

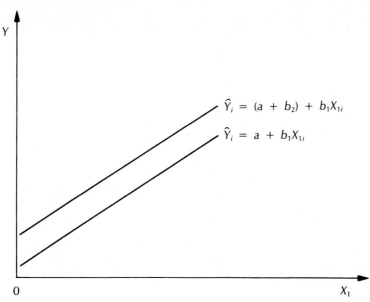

Figure 5.5
Model III—Interaction effect due to gender

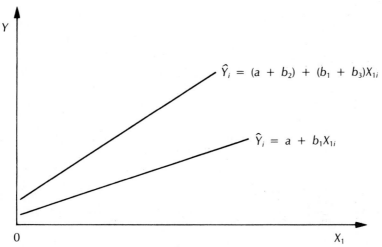

The first two comparisons (II versus I, and III versus II) are straightforward and require only a t-test. The third comparison is new. It is the simultaneous test of two parameters in model III. The test is accomplished by comparing the joint explanatory power of the two predictor variables, X_{2i} and X_{3i}, with the unexplained variation of model III. We compute this explanatory power by comparing the unadjusted R^2 values for models III and I. Thus, the F-statistic for the *incremental* explanatory power of these two predictor variables is

$$F = \frac{(R_{III}^2 - R_I^2)/(DF_I - DF_{III})}{(1 - R_{III}^2)/DF_{III}}$$

where DF_I is the degrees of freedom left for model I

DF_{III} is the degrees of freedom left for model III

In our example, $DF_I = n - 2$, and $DF_{III} = n - 4$.

The intuitive explanation for this F-test is very similar to the F-test for the equation as a whole. The numerator captures the explanatory power for, in this case, a subset of the predictor variables. This incremental explanatory power is divided by the number of degrees of freedom used to obtain higher explanatory power. The denominator defines the unexplained variation for the full model divided by the degrees of freedom left.

Of course, we conduct such an incremental F-test, just like the t-test, only after we have tested the equation as a whole. Thus, we may proceed by first testing the null hypothesis (for model III)

$$H_o : \beta_1 = \beta_2 = \beta_3$$

with

$$F = \frac{R_{III}^2/m}{(1 - R_{III}^2)/(n - m - 1)}$$

where $m = 3$.

If this null hypothesis is rejected, we may choose to conduct model comparisons, as described above.

Problems

Suppose that starting salary is a function of grade-point average and functional classification. Write out the estimated equation, and define all predictor variables for the cases described in Problems 5.1 and 5.2.

5.1 Assume that the five functional classifications shift the straight relationship between starting salary and grade-point average without changing the slope (i.e., model II).

5.2 Assume that the five functional classifications shift the straight relationship between starting salary and grade-point average *and* change the slope (i.e., model III).

5.3 Define an F-test to determine the statistical significance of the intercept differences (i.e., model II versus model I), based on incremental R^2.

5.4 Define an F-test to determine the statistical significance of the slope differences (i.e., model III versus model II), based on incremental R^2.

5.5 Define an F-test to determine the joint statistical significance of the intercept and slope differences (i.e., model III versus model I), based on incremental R^2.

Interaction versus Correlation

It is important to distinguish between the concepts of interaction and correlation. For any pair of variables we can determine a correlation coefficient. This coefficient defines the degree of linear association between the two variables. An *interaction effect* exists if the value of the slope coefficient for one predictor variable depends on the level of another predictor variable. It turns out that our ability to obtain reliable estimates of interaction effects is the highest when the relevant predictor variables are not correlated. Yet in some applications interactions are especially likely to exist if there is substantial correlation. For example, if the price elasticity of a product depends on the level of advertising, a change in advertising should be accompanied by a change in the price, resulting in a correlation between these variables. In general, our understanding and knowledge should provide guidance regarding the possible existence of interaction effects. And given the multitude of reasons for (high) correlations between predictor variables, we should not use their magnitudes as a basis for considering interaction effects.

An Indicator Variable to Distinguish War and Peace Periods in Economic Data

Not only can indicator variables be used to include categorical variables in regression analysis, they can also be used to allow for other differences in the relationship. For example, suppose that we have data over the period 1935 to 1965 for a study of the relation between consumption and income. The sample period includes the Second

World War. Now suppose that during the war the effect of income on consumption differs from the effect in other years. One possibility is to simply ignore the war years and not use the corresponding data. Alternatively, we could incorporate our belief of a difference in the effect into the model by using an indicator variable. The basic model of interest is

$$Y_t = \alpha + \beta X_t + u_t$$

where $Y_t =$ per capita personal consumption expenditures (in 1958 dollars) in year t and

$X_t =$ per capita personal disposable income (in 1958 dollars) in year t.

Let the war years be the years 1939 through 1945. Then we define the following indicator variable as

$D_t = 1$ if the observation represents any of the years 1939 through 1945

$= 0$ otherwise

The model then becomes

$$Y_t = \alpha + \beta_1 X_t + \beta_2 D_t + u_t$$

This model implies that the intercept differs between war and peace years. In other words,

$$Y_t = \alpha + \beta_1 X_t + u_t \qquad \text{for peace years}$$

$$Y_t = (\alpha + \beta_2) + \beta_1 X_t + u_t \qquad \text{for war years}$$

It is conceivable, however, that the war's effects could also change the slope coefficient. To allow for slope differences we add the following variable:

$$(XD)_t = X_t * D_t$$

The full model then is

$$Y_t = \alpha + \beta_1 X_t + \beta_2 D_t + \beta_3 (XD)_t + u_t$$

It is easy to see that for observations taken during the war, $(XD)_t$ will equal X_t, and for all other observations $(XD)_t$ will equal zero, since D_t equals one during the war and zero in all other years. This model implies that both the intercepts *and* the slope parameters may differ. In other words,

$$Y_t = \alpha + \beta_1 X_t + u_t \qquad \text{for peace years}$$

$$Y_t = (\alpha + \beta_2) + (\beta_1 + \beta_3) X_t + u_t \qquad \text{for war years}$$

Thus, β_2 is the difference in intercepts between the two time periods and β_3 is the difference in slope parameters.

Indicator Variables for Seasonal Differences

Suppose we are interested in a model that explains and predicts quarterly per capita consumption of shrimp in the United States. After considerable reflection on the factors that may cause variation in shrimp consumption per capita, we decide that the following model should be tested and estimated:

$$C_t = \alpha + \beta_1 P_t + \beta_2 I_t + \beta_3 X_{2t} + \beta_4 X_{3t} + \beta_5 X_{4t}$$
$$+ \beta_6 Z_{2t} + \beta_7 Z_{3t} + \beta_8 Z_{4t} + u_t$$

where C_t is per capita consumption of shrimp in quarter t

P_t is wholesale price of shrimp in quarter t

I_t is per capita income in thousands of dollars in quarter t

$X_{2t} = 1$ if the observation is in the second quarter of a given year

$\quad\quad = 0$ otherwise

$X_{3t} = 1$ if the observation is in the third quarter of a given year

$\quad\quad = 0$ otherwise

$X_{4t} = 1$ if the observation is in the fourth quarter of a given year

$\quad\quad = 0$ otherwise

$Z_{2t} = I_t * X_{2t}$

$Z_{3t} = I_t * X_{3t}$

$Z_{4t} = I_t * X_{4t}$

The model allows for a linear effect of price, a linear effect of income, differences in the intercepts between the quarters, and differences in the income effects between the quarters. Each effect or difference will be estimated simultaneously by holding the other predictor variables in the equation constant.

For the purpose of this application, we assume that this model is appropriate for testing and estimating, and that the assumptions about the error term are satisfied. For example, we assume that supply of shrimp is not a factor. We concentrate, therefore, on (1) whether the model as a whole is statistically significant, and (2) whether the model can be simplified, that is, whether the model includes irrelevant predictor variables. In practice, we need to investigate the validity of the error assumptions (see Chapter 7). For example, we may question whether the effects of price and income are in fact linear (see Chapter 6).

A Test of the Model as a Whole

Our first test is to determine whether the estimated model has significant explanatory power. The null hypothesis is $H_o : \beta_1 = \beta_2 = \ldots = \beta_8$. Under this null hypothesis, the ratio

$$\frac{R^2/m}{(1 - R^2)/(n - m - 1)}$$

follows the F-distribution with m and $(n - m - 1)$ degrees of freedom. The multiple regression results are shown below for the period 1956–67[3] (i.e., $n = 48$).

Predictor variable	Regression coefficient	Estimated standard error	Computed t-value
Intercept	0.0621	0.0513	1.21
Price (P_t)	−0.1448	0.0326	4.43
Income (I_t)	0.1670	0.0262	6.38
Intercept X_{2t}	−0.0056	0.0716	0.08
Dummies X_{3t}	0.0383	0.0708	0.54
X_{4t}	0.0372	0.0697	0.53
Income Z_{2t}	0.0038	0.0339	0.11
Slope Z_{3t}	−0.0001	0.0333	0.00
Dummies Z_{4t}	0.0008	0.0326	0.03

	Unadjusted	Adjusted
R^2	0.8282	0.80

The results suggest that only two slope coefficients are statistically significant (based on the t-values). However, before we make conclusions about the relevance of the individual predictor variables, we must first perform the F-test for the model as a whole:

$$F = \frac{0.8282/8}{0.1718/39} = 23.5$$

At the five-percent level, the critical F-value equals 2.18 ($F_{8,40}$). Thus, we can reject the null hypothesis. At least one of the eight parameters is different from zero.

We can now test hypotheses about individual parameters in the model. However, one factor that complicates this is the amount of multicollinearity that may exist among the predictor variables. For that reason we do not eliminate all X and Z variables simultaneously, based on the low t-values for all these predictor variables. Instead, we often eliminate one predictor variable at a time and then re-estimate the simplified equation. If the multicollinearity is reduced, the estimated standard errors for some of the remaining predictor variables may decrease.

In this application, it seems more appropriate, however, to consider eliminating *subgroups* of predictor variables. Specifically, X_2 through X_4 allow for differences in the intercepts, while Z_2 through Z_4 allow for differences in the income effects. Thus, the Z variables may be evaluated as a group. For example, we may test the null hypothesis

$$H_o : \beta_6 = \beta_7 = \beta_8 = 0$$

Conceptually, the test of this hypothesis involves the notion of the explanatory power of the Z variables as a group—that is, the explanatory power of the *full* model (with eight predictor variables) compared with the explanatory power of a *restricted* model (without the Z variables). For this we can use an F-test, based on incremental R^2:

$$F = \frac{(R_F^2 - R_R^2)/(DF_R - DF_F)}{(1 - R_F^2)/DF_F}$$

where R_F^2 is the unadjusted R^2 for the full model

R_R^2 is the unadjusted R^2 for the restricted model

DF_F is the number of degrees of freedom left for the full model

DF_R is the number of degrees of freedom left for the restricted model

This F-test is the ratio of the average explanatory power of the three Z variables and the mean squared residual for the full model.

If the comparison of the full and restricted models involves only one predictor variable, this test is

$$F = \frac{(R_F^2 - R_R^2)/1}{(1 - R_F^2)/(n - m - 1)}$$

It can be shown that this F-test equals the square of the t-test for the predictor variable in question, and that it provides identical conclusions.

The multiple regression results for the restricted model are shown below.

Predictor variable	Regression coefficient	Estimated standard error	Computed t-value
Intercept	0.0597	0.0259	2.31
Price (P_t)	−0.1445	0.0312	4.63
Income (I_t)	0.1680	0.0141	11.89
Intercept X_{2t}	0.0023	0.0098	0.24
Dummies X_{3t}	0.0382	0.0098	3.89
X_{4t}	0.0389	0.0098	3.95

	Unadjusted	Adjusted
R^2	0.8281	0.81

The incremental F-test result is

$$F = \frac{(0.8282 - 0.8281)/3}{(0.1718)/39} = 0.01$$

Clearly, we cannot reject the null hypothesis that all three parameters for the Z variables are zero. We conclude that these predictor variables are irrelevant, and that we do not need to allow for differences in the income effects between the four quarters.

Evidence of Multicollinearity

From the results for the restricted model we also note that the computed t-values for the intercept dummy coefficients are much higher than they are in the full model. Indeed, a comparison between the two models of the estimated standard errors for the three X variables suggests the existence of substantial multicollinearity in the full model. For example, the ratio of the estimated standard errors for X_2 (in the full relative to the restricted model) is about 7.3. This increase in estimated standard error is consistent with a simple correlation coefficient for two predictor variables of 0.99 (see Table 4.2).

Source of Multicollinearity

The reason for the high degree of multicollinearity is the use of the X and Z variables together in the same equation. We show below a partial table of entries for these variables.

Period	X_2	X_3	X_4	Z_2	Z_3	Z_4
1	0	0	0	0	0	0
2	1	0	0	I_2	0	0
3	0	1	0	0	I_3	0
4	0	0	1	0	0	I_4
5	0	0	0	0	0	0
6	1	0	0	I_6	0	0
7	0	1	0	0	I_7	0
8	0	0	1	0	0	I_8

We can see from this partial table that X_2 and Z_2 have many entries in common. Thus, X_2 and Z_2 are *necessarily* positively correlated (given that $I_t > 0$ for all t). Indeed, for the sample period the correlation between X_2 and Z_2 equals 0.9805. Similarly, X_3 and Z_3 are highly correlated (0.9856) and so are X_4 and Z_4 (0.9849). If the income effect depends on the quarter, we need substantially more information (for example, a much larger sample size) to be able to obtain reliable differences in the income effects between the four quarters.

Further Simplification

If our choice is between the full and restricted models, the incremental F-test result makes it clear that we should choose the restricted model. An inspection of the computed t-values for individual coefficients of the restricted model suggests that the seasonal differences in the intercepts can be simplified further. For example, the small t-ratio for X_2 suggests that there is no evidence of a difference in the intercepts between the second and first quarter. However, both the third and the fourth quarter are statistically significantly different from the first. Yet the magnitudes of these differences are about the same (0.0382 and 0.0389). Thus, we propose that the final model may be

$$\hat{C}_t = a + b_1 P_t + b_2 I_t + b_3 D_t$$

where $D_t = 1$ if the observation is in the third or fourth quarter
  $= 0$ otherwise

An F-test for the Final Model?

We could also do an F-test for the final model, to test the null hypothesis

$$H_o: \beta_1 = \beta_2 = \beta_3 = 0$$

However, strictly speaking this test is not valid, given that it would be performed after several iterations (i.e., the elimination of the Z variables, and the simplication of the X variables into one D variable). Specifically, this F-test is guaranteed to result in a rejection of the null hypothesis shown above, given that (1) the F-test for the full model (eight predictor variables) showed that at least one of these predictors is relevant and (2) we eliminated those predictors for which the incremental explanatory power is negligible. Thus, as we have argued before, there is no reason to perform an F-test on the final model.

An F-test for the Restricted Model

We could also have performed an F-test for the restricted model. That is, $H_o: \beta_1 = \beta_2 = \beta_3 = \beta_4 = \beta_5$, where the estimated model is

$$\hat{C}_t = a + b_1 P_t + b_2 I_t + b_3 X_{2t} + b_4 X_{3t} + b_5 X_{5t}$$

We obtained an unadjusted R^2 value of 0.8281 for this model. Accordingly,

$$F = \frac{R^2/5}{(1 - R^2)/42} = \frac{0.8281/5}{0.1719/42} = 40.5$$

We may ask ourselves whether we could not have rejected the full model, given that the computed F value of 40.5 exceeds the computed F value for the full model of 23.5. Unfortunately, this comparison is not appropriate (although the conclusion in favor of the restricted model would have been consistent with the result of the incremental R^2 test). The problem below illustrates this by example.

Problem

A researcher has estimated two models for the same set of data. One model (the full model) contains ten predictor variables. The other model (the restricted model) has eight predictor variables, a subset of the ten predictors in the full model. The researcher is unsure about which model to favor. Some relevant results are shown below.

Model	Sample size	Number of predictors	Unadjusted R^2	Computed F-value
Full	100	10	0.60	13.33
Restricted	100	8	0.57	15.06

Assuming that you have to decide which of the two models to favor (there are no other options), which model would you choose? Why?

Answer

Given that the restricted model has a subset of the ten predictors in the full model, we can proceed to test the following null hypothesis:

$$H_o: \beta_9 = \beta_{10} = 0$$

where β_9 and β_{10} are the parameters for the two predictors that appear in the full model but not in the restricted model.

$$F = \frac{(R_F^2 - R_R^2)/(DF_R - DF_F)}{(1 - R_F^2)/DF_F}$$

$$= \frac{(0.60 - 0.57)/2}{(1 - 0.60)/(100 - 10 - 1)}$$

$$= \frac{0.03/2}{0.40/89} = 3.34$$

The tabulated value for a five-percent level of significance equals $F_{2,60} = 3.15$. Thus, we can reject the null hypothesis that the two predictors have no incremental explanatory power. We conclude that the restricted model should be rejected in favor of the full model, even though the computed F-value for the overall test of the full model is lower than the computed F-value for the overall test of the restricted model.

Specification Error

If there are truly m relevant predictor variables in the model and one or more of these variables is omitted, the parameter estimates of the included variables will in general be biased. Suppose, for example, that we have included two predictor variables in the model, X_1 and X_2, but we have omitted two other relevant variables, X_3 and X_4. Then the multiple regression estimator b_1 (see the formula in Chapter 4) only holds X_2 constant, not X_3, or X_4. This estimator is biased unless X_1 is uncorrelated with both X_3 and X_4. Furthermore, X_2 must be uncorrelated with both X_3 and X_4 (unless X_1 and X_2 are uncorrelated). Analogous statements apply to the estimator b_2.

In practice, it is difficult to determine the correct model; however, it is critically important that we appreciate the consequences of omitting relevant variables from the model. Only knowledge and insight about the problem being analyzed will allow us to specify the relevant predictor variables.

SUMMARY

In this chapter we have introduced and discussed the general, linear model. The aspects considered here apply to a model with any number, m, of predictor variables. We have generalized the problem of extreme multicollinearity to the case of more than two predictor variables. And we have suggested how less-than-extreme multicollinearity can be detected as well as, perhaps, resolved.

For the general, linear model the first statistical test should be a test of the model as a whole. Only if the model has significant explanatory power does it make sense to evaluate the significance of individual slope coefficients. In addition, we have to consider all factors that may be responsible for the insignificance of a particular slope coefficient: improper functional form, omitted relevant variables, multicollinearity, and irrelevance of the variable in question.

A large part of the chapter consists of a discussion about indicator variables. We have provided a detailed treatment, including the specification and interpretation of indicator variables using any number of categories. In addition, we have shown how interaction effects can be accommodated and tested.

In practice, there is often confusion about the difference between correlation and interaction. An important distinction is that interactions involve complications in the effects, while correlations between predictor variables are measures of the difficulties we may face if we attempt to accommodate interactions. In an application involving seasonal indicator variables, multicollinearity is a factor contributing to the inability of identifying significant interaction effects.

ADDITIONAL PROBLEMS

5.6 At a graduate school of business all MBA students were required to take a data analysis course. The results of a midterm examination were recorded, and an attempt was made to predict the students' scores based on certain background characteristics: gender, undergraduate major, calculus background, prior statistics course work, and work experience involving statistics. Complete data were available for 113 students.

The variables are defined as follows:

SCORE = Score on the midterm examination

CALC = 1 if student had exposure to calculus
       = 0 otherwise

STAT = 1 if student had prior course work in statistics
       = 0 otherwise

EXPER = 1 if student had work experience in statistics
       = 0 otherwise

SEX   = 1 if female
       = 0 if male

HUM   = 1 if humanity or behavioral sciences undergraduate
       = 0 otherwise

MATH = 1 if mathematics or physical sciences undergraduate
       = 0 otherwise

ECON = 1 if economics or business undergraduate
       = 0 otherwise

Note: the students who had none of these undergraduate backgrounds were all classified as engineering undergraduates

Use the computer output below to answer the following questions.

a. Is there any empirical evidence to support the view that exposure to calculus favorably affects exam performance in data analysis? Why or why not?
b. What is the predicted difference in the exam score for a student with prior course work in statistics *and* work experience in statistics compared with a student with neither prior course work nor work experience in statistics?
c. Rank the four undergraduate-major classes in order of performance on the midterm exam, adjusting for the effects of the other predictor variables.

Selected results are printed below.

> MEAN

VARIABLE	MEAN	STD. DEV.
SCORE	8.37699E+01	8.61582E+00
CALC	8.23009E-01	3.83362E-01
STAT	5.92920E-01	4.93479E-01
EXPER	2.03540E-01	4.04424E-01
SEX	2.30089E-01	4.22764E-01
HUM	3.71681E-01	4.85406E-01
MATH	1.23894E-01	3.30928E-01
ECON	3.09735E-01	4.64444E-01

BASED ON 113 ACTIVE ROWS.

> CORR

	SCORE	CALC	STAT	EXPER	SEX	HUM	MATH	ECON
SCORE	1.000							
CALC	0.223	1.000						
STAT	0.337	0.371	1.000					
EXPER	0.349	0.177	0.419	1.000				
SEX	−0.086	−0.132	0.153	−0.015	1.000			
HUM	−0.216	−0.411	−0.444	−0.161	0.189	1.000		
MATH	0.045	0.174	−0.016	−0.057	−0.078	−0.289	1.000	
ECON	−0.031	0.110	0.360	−0.006	0.043	−0.515	−0.252	1.000

Criterion Variable: SCORE

> COEF

VARIABLE	B	STD. ERROR(B)	T
CALC	1.0963E+00	2.2430E+00	0.489
STAT	4.3636E+00	2.0203E+00	2.160
EXPER	4.3047E+00	2.1001E+00	2.050
SEX	−.1571E+01	1.9046E+00	−0.825
HUM	−.3065E+01	2.3985E+00	−1.278
MATH	−.1599E+01	2.7568E+00	−0.580
ECON	−.4201E+01	2.26224+00	−1.857
CONSTANT	8.2404E+01	2.8371E+00	29.046

> SUMM

	MULTIPLE R	R−SQUARE
UNADJUSTED	0.4625	0.2139
ADJUSTED	0.4019	0.1615

STD. DEV. OF RESIDUALS =   7.88943

Criterion Variable: SCORE

```
>COEF

VARIABLE        B         STD. ERROR(B)      T

STAT       4.1960E+00    1.8708E+00       2.243
EXPER      4.6392E+00    2.0537E+00       2.259
ECON       -.3910E+01    1.9223E+00      -2.034
HUM        -.3251E+01    1.8823E+00      -1.727
CONSTANT   8.2757E+01    1.7312E+00      47.804

>SUMM

                MULTIPLE R   R-SQUARE

UNADJUSTED       0.4508      0.2032
ADJUSTED         0.4168      0.1737

STD. DEV. OF RESIDUALS  =     7.83193
```

5.7 Consider the following undergraduate majors for a group of MBA students, and the number of observations for each category.

Major	Number of Observations
Behavioral science	15
Business administration	8
Economics	32
Engineering	26
Humanities	45
No major	3
Other	18

a. Using this information, construct a set of indicator variables to allow for differences in the effect of undergraduate major on the grade-point average for a required course in accounting. Justify your choice of variables, and show your expectations about signs and relative magnitudes of coefficients.

b. Follow the same procedure as in 5.7a, except consider the grade-point average for a required course in marketing.

c. Follow the same procedure as in 5.7a, except consider the criterion variable as a measure of the average grade-point average across five courses: accounting, statistics, marketing, finance, and organizational behavior.

5.8 According to a February 24, 1980, article in the *Wall Street Journal,* MBA graduates from a leading business school earn starting salaries as high as $40,000. Suppose that the recruitment process is set up to preserve the consulting industry's competitive structure. Obviously, MBA students tend to favor companies which make the more lucrative offers. Naturally, the better a company's financial condition, the more money it can offer. To balance the bidding process, the school might institute a draft procedure analogous to existing schemes for professional football organizations. Imagine that five consulting companies were invited to participate in a draft of four rounds for a total of twenty students. The companies were rank-ordered according to profits per employee, and the company with the lowest profits is allowed first choice in each round.

To determine the effects of draft round and other variables on the students' starting salaries, the following model was constructed:

$$Y_i = \alpha + \beta_1 X_{1i} + \beta_2 X_{2i} + \beta_3 X_{3i} + \beta_4 X_{4i}$$
$$+ \beta_5 X_{5i} + \beta_6 X_{6i} + \beta_7 X_{7i} + u_i$$

where Y_i = starting salary in $000 (SALARY)

X_{1i} = 1 if drafted in round 1 (ROUND 1)

    = 0 otherwise

X_{2i} = 1 if drafted in round 2 (ROUND 2)

    = 0 otherwise

X_{3i} = 1 if drafted in round 3 (ROUND 3)

    = 0 otherwise

X_{4i} = height in inches (HEIGHT)

X_{5i} = age in years (AGE)

X_{6i} = average exam score for all decision-science courses (DECSCI)

X_{7i} = average exam score for all accounting courses (ACCTG)

Use the selected computer output to answer the following questions, assuming that the assumptions needed for statistical inference are satisfied.

a. Test the null hypothesis that

$$\beta_1 = \beta_2 = \cdots = \beta_7 = 0$$

against the alternative that at least one of the above β values is different from zero (i.e., determine the statistical significance of the overall multiple regression model).

b. According to the results, what is the average monetary value associated with being drafted in the second round as opposed to the fourth round while holding the other predictor variables constant?

c. As expected, MBA's drafted in the second round have a higher salary on the average than MBA's drafted in the fourth round (i.e., the coefficient for ROUND2 equals 3.2686). Using a type I error of ten percent, test the null hypothesis that there is no difference between the salaries of MBA's drafted in the second and fourth rounds, based on the multiple regression.

d. Based on the simple correlation matrix, ACCTG has a positive correlation with SALARY. Yet, in the multiple regression, ACCTG has a negative coefficient. If we are interested in understanding the true difference resulting from the average score in accounting courses on salary, which result—simple correlation coefficient or multiple regression coefficient—should be used? Why?

e. AGE has the lowest t-ratio of all predictor variables. Is it reasonable to conclude that AGE is not a factor in recruiting by the five consulting companies? Why or why not?

```
> MEAN

VARIABLE        MEAN         STD. DEV.

SALARY       3.50149+01     6.58662E+00
ROUND1       2.50000E-01    4.44262E-01
ROUND2       2.50000E-01    4.44262E-01
ROUND3       2.50000E-01    4.44262E-01
HEIGHT       6.89001E+01    3.52607E+00
AGE          2.94615E+01    2.91537E+00
DECSCI       8.35000E+01    8.35716E+00
ACCTG        6.95884E+01    1.00850E+01

BASED ON 20 ACTIVE ROWS.
```

```
> CORR

         SALARY  ROUND1  ROUND2  ROUND3  HEIGHT   AGE   DECSCI  ACCTG
SALARY    1.00
ROUND1    0.12    1.00
ROUND2   -0.04   -0.33
ROUND3   -0.22   -0.33   -0.33    1.00
HEIGHT    0.71   -0.27   -0.23    0.19    1.00
AGE      -0.08   -0.01   -0.18    0.04    0.25    1.00
DECSCI    0.38   -0.19    0.35   -0.26   -0.01   -0.51   1.00
ACCTG     0.20   -0.05    0.24    0.02    0.00   -0.22   0.76    1.00
```

```
> COEF

VARIABLE        B         STD. ERROR(B)       T

ROUND1      8.7940E+00     2.1614E+00       4.069
ROUND2      3.2686E+00     1.9154E+00       1.706
ROUND3      1.4197E+00     2.2833E+00       0.622
HEIGHT      1.6740E+00     1.9925E-01       8.402
AGE         1.1998E-01     3.0804E-01       0.390
DECSCI      6.3027E-01     1.9410E-01       3.247
ACCTG      -.2755E+00      1.2951E-01      -2.127
CONSTANT   -.1207E+03      2.0028E+01      -6.026
```

```
> SUMM       MULTIPLE R   R-SQUARE

UNADJUSTED      0.9471       0.8969
ADJUSTED        0.9148       0.8368

STD. DEV. OF RESIDUALS = 2.6611

> ANOV

SOURCE            SS        DF      MS         F

REGRESSION    7.39308E+02    7   1.05615E+02   14.91
RESIDUALS     8.49782E+01   12   7.08152E+00
TOTAL         8.24287E+02   19   4.33835E+01
```

APPENDIX 5.1
SPECIFICATION ERROR (OMITTED VARIABLES)

Suppose that the true model is

$$Y_i = \alpha + \beta_1 X_{1i} + \beta_2 X_{2i} + \beta_3 X_{3i} + \beta_4 X_{4i} + u_i$$

But instead we use

$$Y_i = \alpha + \beta_1 X_{1i} + \beta_2 X_{2i} + v_i$$

Then the estimator b_1, based on the incorrect model, is as follows (variables are in deviation form):

$$b_1 = \frac{(\Sigma y_i x_{1i})(\Sigma x_{2i}^2) - (\Sigma y_i x_{2i})(\Sigma x_{1i} x_{2i})}{(\Sigma x_{1i}^2)(\Sigma x_{2i}^2) - (\Sigma x_{1i} x_{2i})^2}$$

And, it can be shown that

$$E(b_1) = \beta_1 + \beta_3 \frac{\Sigma x_{2i}^2 \Sigma x_{1i} x_{3i} - \Sigma x_{1i} x_{2i} \Sigma x_{2i} x_{3i}}{\Sigma x_{1i}^2 \Sigma x_{2i}^2 - (\Sigma x_{1i} x_{2i})^2}$$

$$+ \beta_4 \frac{\Sigma x_{2i}^2 \Sigma x_{1i} x_{4i} - \Sigma x_{1i} x_{2i} \Sigma x_{2i} x_{4i}}{\Sigma x_{1i}^2 \Sigma x_{2i}^2 - (\Sigma x_{1i} x_{2i})^2}$$

Thus, the bias is a function of the covariance between X_1 and X_3, X_1 and X_2, X_2 and X_3, X_1 and X_4, and X_2 and X_4, as well as the magnitudes of the parameters for the omitted variables.

APPENDIX 5.2
THE GENERAL LINEAR MODEL IN MATRIX NOTATION

In this appendix we provide a short treatment of the general linear model in matrix form. We assume familiarity with the basics of matrix algebra. For an introduction to the basics, see Appendix 5.3.

The model is

$$Y_i = \beta_1 + \beta_2 X_{2i} + \beta_3 X_{3i} + \cdots + \beta_k X_{ki} + u_i$$

Note that we now use β_1 for the intercept. Also, there are k parameters in the model, but $(k-1)$ predictor variables. In matrix notation, this model is

$$y = X\beta + u$$

where

$$y = \begin{bmatrix} Y_1 \\ Y_2 \\ \vdots \\ Y_n \end{bmatrix} \qquad X = \begin{bmatrix} 1 & X_{21} & \cdots & X_{k1} \\ 1 & X_{22} & & X_{k2} \\ \vdots & \vdots & \ddots & \vdots \\ 1 & X_{2n} & \cdots & X_{kn} \end{bmatrix}$$

$$\beta = \begin{bmatrix} \beta_1 \\ \beta_2 \\ \vdots \\ \beta_k \end{bmatrix} \qquad u = \begin{bmatrix} u_1 \\ u_2 \\ \vdots \\ u_n \end{bmatrix}$$

In terms of the dimensions of the vectors and matrices, we have

$$\begin{array}{cccc} y & = & X\beta & + & u \\ (n \times 1) & & (n \times k)(k \times 1) & & (n \times 1) \end{array}$$

The estimated model is

$$\hat{y} = X\hat{\beta}$$

and

$$\hat{u} = y - \hat{y} = \begin{bmatrix} Y_1 - \hat{Y}_1 \\ Y_2 - \hat{Y}_2 \\ \vdots \\ Y_n - \hat{Y}_n \end{bmatrix} = \begin{bmatrix} \hat{u}_1 \\ \hat{u}_2 \\ \vdots \\ \hat{u}_n \end{bmatrix}$$

As before, we desire estimates, $\hat{\beta}$, such that the sum of the squared residuals is minimized. That is, we want to minimize $\Sigma \hat{u}_i^2$, or

Minimize $\hat{u}'\hat{u}$

or

Minimize $(y - \hat{y})'(y - \hat{y}) = (y - X\hat{\beta})'(y - X\hat{\beta})$

That is, we choose $\hat{\beta}$ so that the quadratic form shown above is minimized. It is easy to show that

$$\frac{\partial \hat{u}'\hat{u}}{\partial \hat{\beta}} = -2X'y + 2X'X\hat{\beta}$$

Minimization may be achieved by setting the partial derivatives equal to zero, which results in

$$X'X\hat{\beta} = X'y$$

From this we can obtain an expression for $\hat{\beta}$, as long as $X'X$ has full rank and $(X'X)^{-1}$ exists (or the determinant of $X'X$ is nonzero). This is equivalent to requiring that the system of equations (for the intercept and the slope coefficients) is linearly independent. Then,

$$\hat{\beta} = (X'X)^{-1} X'y$$

To check whether a minimum is achieved, we can take the second partial derivatives:

$$\frac{\partial^2 (\hat{u}'\hat{u})}{\partial \hat{\beta}^2} = 2X'X$$

Given that $X'X$ is positive definite (each entry on the diagonal of this matrix represents the sum of the squared values for a predictor variable), $\hat{\beta}$ does indeed minimize the sum of squared residuals.

With appropriate assumptions about the error term in the model, we can show that the estimator is unbiased. Additional assumptions allow us to obtain an expression for the variance–covariance matrix of $\hat{\beta}$. For unbiasedness, we require

$$E(u) = 0$$

To show that the estimator is unbiased under this assumption, we also require that the predictor variables represent a set of fixed numbers. Together these assumptions are violated if there are omitted variables which are correlated with included predictors, or if the functional form is incorrect.

To show unbiasedness

$$\hat{\beta} = (X'X)^{-1}X'y$$
$$= (X'X)^{-1}X'(X\beta + u)$$
$$= \beta + (X'X)^{-1}X'u$$

$$E(\hat{\beta}) = \beta + (X'X)^{-1}X'E(u) \quad \text{if } X \text{ is fixed}$$
$$= \beta \quad \text{if } E(u) = 0$$

The variance–covariance matrix of $\hat{\beta}$ is obtained by assuming

$$E(uu') = \sigma^2 I_n$$

where

$$uu' = \begin{bmatrix} u_1^2 & u_1u_2 & \cdots & u_1u_n \\ u_2u_1 & u_2^2 & \cdots & u_2u_n \\ \vdots & \vdots & \ddots & \vdots \\ u_nu_1 & u_nu_2 & \cdots & u_n^2 \end{bmatrix} \quad \text{and}$$

$$E(uu') = \begin{bmatrix} E(u_1^2) & E(u_1u_2) & \cdots & E(u_1u_n) \\ E(u_2u_1) & E(u_2^2) & \cdots & E(u_2u_n) \\ \vdots & \vdots & \ddots & \vdots \\ E(u_nu_1) & E(u_nu_2) & \cdots & E(u_n^2) \end{bmatrix}$$

By assumption, this matrix is

$$E(uu') = \begin{bmatrix} \sigma^2 & 0 & \cdots & 0 \\ 0 & \sigma^2 & \cdots & 0 \\ \vdots & \vdots & \ddots & \vdots \\ 0 & 0 & \cdots & \sigma^2 \end{bmatrix}$$

Note that this matrix incorporates the assumptions of independence and homoscedasticity for the error term. All the diagonal elements are equal (homoscedasticity) and all off-diagonal elements are zero (independence). It is a straightforward process to derive the result

$$\text{var}(\hat{\beta}) = \sigma^2(X'X)^{-1}$$

APPENDIX 5.3
THE BASICS OF MATRIX ALGEBRA

Addition of matrices (or vectors)

For $C = A + B$ to exist, we require that A and B are of equal dimensions. In general, let A be of dimensions $(m \times n)$, then B must be $(m \times n)$ and C will be $(m \times n)$.

Example

$$A = \begin{bmatrix} a_{11} & a_{21} \\ a_{12} & a_{22} \\ a_{13} & a_{23} \end{bmatrix} = \begin{bmatrix} 5 & 12 \\ 7 & 14 \\ 8 & 10 \end{bmatrix}$$

$$B = \begin{bmatrix} b_{11} & b_{21} \\ b_{12} & b_{22} \\ b_{13} & b_{23} \end{bmatrix} = \begin{bmatrix} 4 & -2 \\ 3 & -8 \\ 5 & -7 \end{bmatrix}$$

$$C = \begin{bmatrix} (a_{11} + b_{11}) & (a_{21} + b_{21}) \\ (a_{12} + b_{12}) & (a_{22} + b_{22}) \\ (a_{13} + b_{13}) & (a_{23} + b_{23}) \end{bmatrix}$$

$$= \begin{bmatrix} (5 + 4) & (12 - 2) \\ (7 + 3) & (14 - 8) \\ (8 + 5) & (10 - 7) \end{bmatrix} = \begin{bmatrix} 9 & 10 \\ 10 & 6 \\ 13 & 3 \end{bmatrix}$$

Multiplication of matrices (or vectors)

For $C = AB$ to exist, we require that the number of columns in A equals the number of rows in B, that is, let A be $(m \times n)$, then B must be $(n \times p)$, and C will be $(m \times p)$.

Example

$$A = \begin{bmatrix} a_{11} & a_{21} \\ a_{12} & a_{22} \\ a_{13} & a_{23} \end{bmatrix} = \begin{bmatrix} 5 & 12 \\ 7 & 14 \\ 8 & 10 \end{bmatrix}$$

$$B = \begin{bmatrix} b_{11} & b_{21} \\ b_{12} & b_{22} \end{bmatrix} = \begin{bmatrix} 2 & 3 \\ 5 & 1 \end{bmatrix}$$

$$C = \begin{bmatrix} (a_{11}b_{11} + a_{21}b_{12}) & (a_{11}b_{21} + a_{21}b_{22}) \\ (a_{12}b_{11} + a_{22}b_{12}) & (a_{12}b_{21} + a_{22}b_{22}) \\ (a_{13}b_{11} + a_{23}b_{12}) & (a_{13}b_{21} + a_{23}b_{22}) \end{bmatrix}$$

$$= \begin{bmatrix} (5 \times 2 + 12 \times 5) & (5 \times 3 + 12 \times 1) \\ (7 \times 2 + 14 \times 5) & (7 \times 3 + 14 \times 1) \\ (8 \times 2 + 10 \times 5) & (8 \times 3 + 10 \times 1) \end{bmatrix} = \begin{bmatrix} 70 & 27 \\ 84 & 35 \\ 66 & 34 \end{bmatrix}$$

Note that $c_{ij} = \sum\limits_{k=1}^{n} a_{ik}b_{kj}$

Transpose

The transpose of a matrix A,A', is a matrix that results from interchanging the rows and columns of the matrix A.

Example

$$A = \begin{bmatrix} a_{11} & a_{21} \\ a_{12} & a_{22} \\ a_{13} & a_{23} \end{bmatrix} \quad A' = \begin{bmatrix} a_{11} & a_{12} & a_{13} \\ a_{21} & a_{22} & a_{23} \end{bmatrix}$$

$$A = \begin{bmatrix} 5 & 12 \\ 7 & 14 \\ 8 & 10 \end{bmatrix} \quad A' = \begin{bmatrix} 5 & 7 & 8 \\ 12 & 14 & 10 \end{bmatrix}$$

Determinant

For any *square* matrix A (the numbers of rows equals the number of columns), there exists a quantity (a single value) called the determinant of A, or $|A|$. This quantity is defined as follows for a (2×2) matrix:

$$\begin{vmatrix} a_{11} & a_{21} \\ a_{12} & a_{22} \end{vmatrix} = a_{11}a_{22} - a_{12}a_{21}$$

Example

$$\begin{vmatrix} 25 & 10 \\ 10 & 40 \end{vmatrix} = (25)(40) - (10)(10) = 900$$

For a (3×3) matrix A, the determinant equals

$$|A| = a_{11} \begin{vmatrix} a_{22} & a_{23} \\ a_{32} & a_{33} \end{vmatrix} - a_{12} \begin{vmatrix} a_{21} & a_{23} \\ a_{31} & a_{33} \end{vmatrix} + a_{13} \begin{vmatrix} a_{21} & a_{22} \\ a_{31} & a_{32} \end{vmatrix}$$

$$= a_{11}a_{22}a_{33} - a_{11}a_{23}a_{32} - a_{12}a_{21}a_{33} + a_{12}a_{23}a_{31} + a_{13}a_{21}a_{32} - a_{13}a_{22}a_{31}$$

In general,

$$|A| = \sum_{j=1}^{n} a_{ij}c_{ij} = \sum_{i=1}^{n} a_{ij}c_{ij}$$

where A is $(n \times n)$

i represents the ith row

j represents the jth column

a_{ij} is the element in the ith row and jth column of A

c_{ij} is a co-factor (defined below)

Co-factor

The co-factor $c_{ij} = (-1)^{i+j}|A_{ij}|$
where $|A_{ij}|$ is the determinant of A after row i and column j have been deleted. For a (3×3) matrix A, the determinant equals

$$|A| = \sum_{j=1}^{3} a_{ij}c_{ij} = a_{11}c_{11} + a_{12}c_{12} + a_{13}c_{13}$$

$$= a_{11}(-1)^{1+1}|A_{11}| + a_{12}(-1)^{1+2}|A_{12}| + a_{13}(-1)^{1+3}|A_{13}|$$

$$= a_{11}|A_{11}| - a_{12}|A_{12}| + a_{13}|A_{13}|$$

$$= a_{11} \begin{vmatrix} a_{22} & a_{23} \\ a_{32} & a_{33} \end{vmatrix} - a_{12} \begin{vmatrix} a_{21} & a_{23} \\ a_{31} & a_{33} \end{vmatrix} + a_{13} \begin{vmatrix} a_{21} & a_{22} \\ a_{31} & a_{32} \end{vmatrix}$$

$$= a_{11}a_{22}a_{33} - a_{11}a_{23}a_{32} - a_{12}a_{21}a_{33} + a_{12}a_{23}a_{31}$$
$$+ a_{13}a_{21}a_{32} - a_{13}a_{22}a_{31}$$

Properties of determinants

1. $|A'| = |A|$
2. Interchanging any two rows (or any two columns) of A changes only the sign of $|A|$
3. $|A| = 0$ if A has two identical rows (or two identical columns)

Identity Matrix

An identity matrix I_n is a square matrix with entries as shown below, where n is the number of rows (and columns).

$$I = \begin{bmatrix} 1 & 0 & \cdots & 0 \\ 0 & 1 & \cdots & 0 \\ \vdots & \vdots & \ddots & \vdots \\ 0 & 0 & & 1 \end{bmatrix}$$

The identity matrix has the property that pre- or postmultiplying a matrix A by the identity matrix results in the same matrix A. For example,

$$\begin{array}{ccc} A & I_n = & A \\ (m \times n) & & (m \times n) \end{array}$$

Inverse

The inverse of a matrix A, A^{-1}, if it exists, is defined such that $AA^{-1} = I$. It can be shown that

$$A^{-1} = \frac{1}{|A|} \ (\text{adj } A)$$

where $|A|$ is the determinant of A and adj A is the adjoint of A. The adjoint of A consists of the co-factors c_{ij} transposed. Thus,

$$\text{adj } A = \begin{bmatrix} c_{11} & c_{21} & \cdots & c_{n1} \\ c_{12} & c_{22} & \cdots & c_{n2} \\ \vdots & \vdots & \ddots & \vdots \\ c_{1n} & c_{2n} & \cdots & c_{nn} \end{bmatrix}'$$

Note that because $A^{-1} = \frac{1}{|A|}(\text{adj } A)$, the inverse exists only if the determinant is nonzero.

Example

$$A = \begin{bmatrix} 5 & 15 \\ 5 & 55 \end{bmatrix}$$

$$|A| = a_{11}a_{22} - a_{12}a_{21} = (5)(55) - (15)(15) = 50$$

Since $|A| \neq 0$, A^{-1} exists.

$$\text{Adj } A = \begin{bmatrix} c_{11} & c_{21} \\ c_{12} & c_{22} \end{bmatrix}' = \begin{bmatrix} 55 & -15 \\ -15 & 5 \end{bmatrix}$$

$$A^{-1} = \frac{1}{50} \begin{bmatrix} 55 & -15 \\ -15 & 5 \end{bmatrix} = \begin{bmatrix} \dfrac{55}{50} & \dfrac{-15}{50} \\ \dfrac{-15}{50} & \dfrac{5}{50} \end{bmatrix}$$

Then $AA^{-1} = I$

$$\begin{bmatrix} 5 & 15 \\ 5 & 55 \end{bmatrix} \begin{bmatrix} \dfrac{55}{50} & \dfrac{-15}{50} \\ \dfrac{-15}{50} & \dfrac{5}{50} \end{bmatrix} = \begin{bmatrix} 1 & 0 \\ 0 & 1 \end{bmatrix}$$

Example

$$A = \begin{bmatrix} 5 & 15 & 18 \\ 15 & 55 & 54 \\ 18 & 54 & 76 \end{bmatrix}$$

$$|A| = a_{11} \begin{vmatrix} a_{22} & a_{23} \\ a_{32} & a_{33} \end{vmatrix} - a_{12} \begin{vmatrix} a_{21} & a_{23} \\ a_{31} & a_{33} \end{vmatrix} + a_{13} \begin{vmatrix} a_{21} & a_{22} \\ a_{31} & a_{32} \end{vmatrix}$$

$$= 5(55 \times 76 - 54^2) - 15(15 \times 76 - 18 \times 54) + 18(15 \times 54 - 18 \times 55)$$
$$= 5 \times 1264 - 15 \times 168 + 18 \times (-280) = 760$$

Since $|A| \neq 0$, A^{-1} exists.

$$\text{Adj } A = \begin{bmatrix} c_{11} & c_{21} & c_{31} \\ c_{12} & c_{22} & c_{32} \\ c_{13} & c_{23} & c_{33} \end{bmatrix}'$$

$$c_{11} = |A_{11}| = 55 \times 76 - 54^2 = 1264$$

$$c_{21} = -|A_{21}| = -(15 \times 76 - 18 \times 54) = -168$$

$$c_{31} = |A_{31}| = 15 \times 54 - 18 \times 55 = -280$$

$$c_{12} = -|A_{12}| = -(15 \times 76 - 18 \times 54) = -168$$

$$c_{22} = |A_{22}| = 5 \times 76 - 18^2 = 56$$

$$c_{32} = -|A_{32}| = -(5 \times 54 - 15 \times 18) = -10$$

$$c_{13} = |A_{13}| = 15 \times 54 - 18 \times 55 = -280$$

$$c_{23} = -|A_{23}| = -(5 \times 54 - 15 \times 18) = -10$$

$$c_{33} = |A_{33}| = 5 \times 55 - 15^2 = 50$$

$$A^{-1} = \frac{1}{760} \begin{bmatrix} 1264 & -168 & -280 \\ -168 & 56 & -10 \\ -280 & -10 & 50 \end{bmatrix}$$

Then $AA^{-1} = I$

$$\begin{bmatrix} 5 & 15 & 18 \\ 15 & 55 & 54 \\ 15 & 54 & 76 \end{bmatrix} \begin{bmatrix} \dfrac{1264}{760} & \dfrac{-168}{760} & \dfrac{-280}{760} \\ \dfrac{-168}{760} & \dfrac{56}{760} & \dfrac{-10}{760} \\ \dfrac{-280}{760} & \dfrac{-10}{760} & \dfrac{50}{760} \end{bmatrix} = \begin{bmatrix} 1 & 0 & 0 \\ 0 & 1 & 0 \\ 0 & 0 & 1 \end{bmatrix}$$

ENDNOTES

1. Marquardt, Donald W., and Ronald D. Snee, "Ridge Regression in Practice," *The American Statistician,* 29 (February 1975), 3–19.

2. We continue to refer to the *b*-value as a slope coefficient. However, for an indicator variable this coefficient is the difference between two sample means.

3. Green, Richard D., and John D. Poll, "Dummy Variables and Seasonality—A Curio," *The American Statistician,* 28 (May 1974), 60–62.

REFERENCES

Green, Richard D., and John D. Poll. "Dummy Variables and Seasonality—A Curio." *The American Statistician* 28 (May 1974): 60–62.

Marquardt, Donald W., and Ronald D. Snee. "Ridge Regression in Practice." *The American Statistician* 29 (February 1975): 3–19.

6

Nonlinear Effects

In many empirical studies, only linear effects for the variables of interest are allowed for. Many applications, however, require accommodation of nonlinear effects. In this chapter, we discuss the consequences for prediction and understanding of using an inappropriate functional form (e.g., linear effects instead of nonlinear effects). We show for simple examples that the implications of an inappropriate linear equation can be quite misleading. We propose procedures that can be followed for model specification in terms of the nature of the effects. In all cases, we treat the problem of nonlinear effects through transformations of variables. That is, we continue to use the approach of linear regression analysis, which for transformed variables may be used to accommodate nonlinear effects for the original variables. The resulting model is referred to as being "linear in the parameters."

Few relationships between variables are exactly linear. Thus, the functional form we have assumed and used so far (for a model with one predictor variable),

$$Y_i = \alpha + \beta X_i + u_i$$

is often unrealistic. This model states that a one-unit increase in the value of the predictor variable is expected to increase the value of the criterion variable by β units. No matter what the (current) value for X is, this model implies that we *always* expect Y to increase by β units. It is important, therefore, to ask ourselves if this is a reasonable assumption for the problem we are working on.

Experience Curve

To illustrate the problems that may exist for linear equations, we use an experience-curve example. The literature on experience-curve effects suggests that the (marginal) cost of production tends to decline as the cumulative experience with production increases. For example, as production-department employees continue to produce a certain item, they may discover opportunities to cut costs or ways to enhance the production efficiency. However, the more the marginal cost has been reduced, the more difficult it becomes to reduce it further. Thus, the relationship between the marginal unit cost of production and cumulative experience is often described as follows: "Each time cumulative production doubles, the marginal unit cost of production declines by a constant percentage." Such an experience-curve effect can be shown as follows. Suppose that the marginal cost declines by 30 percent each time cumulative production doubles. Then we could observe the data in the table below (assuming a marginal unit cost of $100 at cumulative production of 1 million units):

An experience curve example

Marginal cost per unit (dollars)	Cumulative production (millions of units)
100	1
70	2
49	4
34	8

Figure 6.1
An experience curve

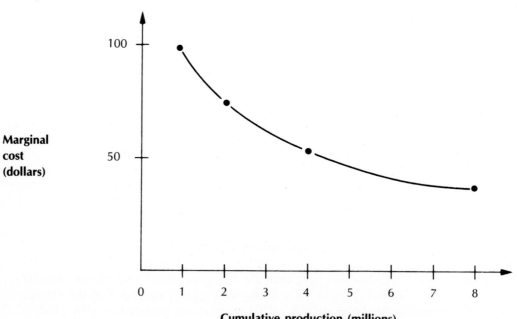

In practice, data would be available for intermediate values of cumulative production, but the data in this table serve to illustrate the idea of an experience curve. It is also easy to see that an assumption of linearity is inappropriate for these data. For example, as production increases from 1 to 2 million, the marginal cost declines by $30. On the other hand, an increase from 2 to 4 million reduces the marginal cost by only $21, or $10.50 per million. The nonlinearity is also evident in Figure 6.1.

It is clear from Figure 6.1 that the slope of the curve is steepest for small values of X, and that this slope slowly approaches zero as X increases. Note also that the slope is changing continuously. That is, there are no two values of X for which the magnitude of the slope is the same. We conclude that it is inappropriate to use a linear equation for the effect of cumulative production on marginal cost. However, if we plot the data on double-logarithmic paper, as shown in Figure 6.2, we see that there is a linear relationship between the logarithmic values (the relationship is "linear in the logarithms").

The logarithmic values are shown in the table below, next to the original values for each observation. For this example we have used the *natural* logarithms (ln) of the values in the table above. The data show that an increase in ln production of about 0.70

Figure 6.2
An experience curve on double-log paper

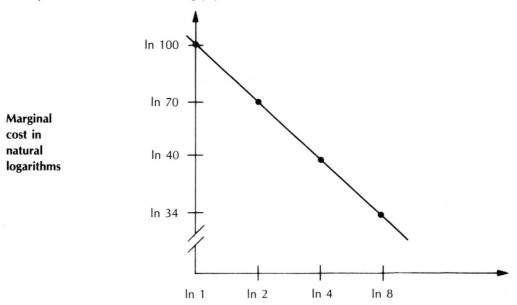

Cumulative production in natural logarithms

Double-log transformations

Marginal unit cost	Ln cost	Ln production	Cumulative production
100	4.61	0	1
70	4.25	0.69	2
49	3.89	1.39	4
34	3.53	2.08	8

corresponds to a decrease in ln cost of 0.36, or a slope coefficient for a linear relationship between ln cost and ln production of about -0.50. This corresponds to a 30 percent decline in marginal unit cost each time cumulative production doubles (see Appendix 6.1).

To summarize, we have an example of a problem where the assumption of linear effects for the original variable is not appropriate. Based on knowledge about the form of the relationship, we search for transformations of the original variables such that there is a linear relation between the transformed variables. In this problem our knowledge is that each time cumulative production doubles, the marginal unit cost declines by a constant percentage.

The relationship between the original variables can be formally stated as follows (we still assume one predictor variable, and include an error term):

$$Y_i = e^\alpha X_i^\beta e^{u_i}$$

where Y_i is marginal unit cost for the ith observation

X_i is cumulative production for the ith observation

u_i is the error associated with the ith observation

e is the base for the system of natural logarithms (e is approximately 2.7183)

By taking natural logarithms of both sides of this equation, we obtain

$$\ln Y_i = \alpha \ln e + \beta \ln X_i + u_i \ln e$$

and, since $\ln e = 1$, this simplifies to

$$\ln Y_i = \alpha + \beta \ln X_i + u_i$$

Thus, we can use the logarithmic values of the original variables and perform a linear regression analysis of the transformed variables. Or, if

$$Y_i^* = \ln Y_i \quad \text{and}$$

$$X_i^* = \ln X_i$$

the model that is linear in the parameters is

$$Y_i^* = \alpha + \beta X_i^* + u_i$$

To show mathematically that the proposed functional form is appropriate, we can also examine the behavior of the first derivative (the slope) of Y with respect to X. We examine this in Appendix 6.2.

Application

Suppose we have obtained the data shown in Table 6.1, relevant to a relationship between marginal cost and cumulative production. Assume that cumulative production is the only relevant predictor variable. In that case we can use the data to verify a particular functional form for the relationship. That is, we can use a scatterplot with Y on the vertical axis and X on the horizontal axis to learn the apparent functional form. However, if there are other relevant predictors, this picture may be misleading (i.e., the other predictor variables should be held constant). In Figure 6.3 we see a scatterplot of the standardized data.[1]

An inspection of the plot confirms the existence of nonlinearity in the relationship between marginal cost and cumulative production. Again, we assume that there are no other relevant predictor variables. Thus, based on our knowledge of experience-curve phenomena, as well as on the scatterplot, we propose a logarithmic transformation for the two variables. The relationship between these transformed variables should be approximately linear.

To verify that the transformations are appropriate, we can inspect scattergrams of the transformed variables. We see in Figure 6.4 that the relationship between the natural logarithm of marginal cost and the natural logarithm of cumulative production is indeed

Table 6.1
An experience-curve application

Y_i Marginal unit cost (dollars)	X_i Cumulative production (millions of units)
80	1
61	2
52	3
46	4
42	5
39	6
37	7
35	8
33	9
32	10

Figure 6.3
Scatterplot of ten standardized values of 'cost' vs. 'production'

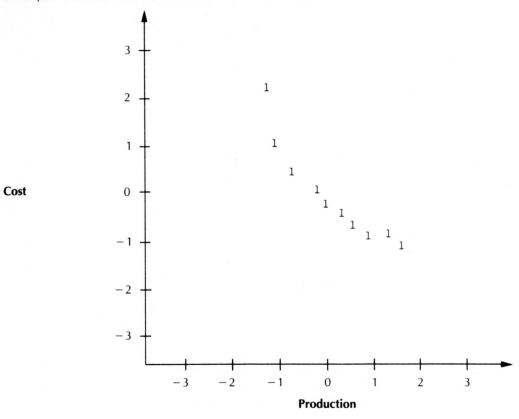

	MEAN	STD. DEV.
VERT. VAR.	45.7	15.0853
HORIZ. VAR.	5.5	3.02765

SAMPLE SIZE = 10

Figure 6.4
Scatterplot of ten standardized values of 'Incost' vs. 'Inprod'

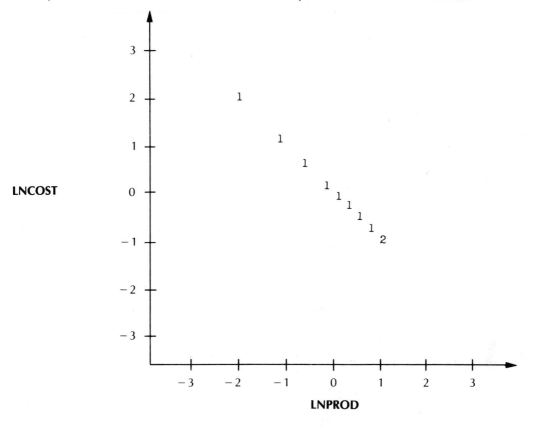

	MEAN	STD. DEV.
VERT. VAR.	3.78025	.293852
HORIZ. VAR.	1.51044	.733024
SAMPLE SIZE = 10		

linear. A regression analysis of the transformed variables provides the following result:

```
> COEF

VARIABLE          B        STD. ERROR(B)         T

LNPROD     -.4008E+00     2.2817E-03    -1.75673E+02
CONSTANT    4.3857E+00    3.7940E-03    +1.15594E+03

> SUMM
                  MULTIPLE R   R-SQUARE

UNADJUSTED      0.9999       0.9997
ADJUSTED        0.9999       0.9997

STD. DEV. OF RESIDUALS = 5.01753E-03
```

We can use the estimated equation to predict the marginal unit cost of production when cumulative production reaches, say, 14 million units. That is,

$$\begin{aligned}
\hat{\text{LNCOST}} &= 4.38 - 0.40 \text{ LNPROD} \\
&= 4.38 - 0.40 \text{ (ln 14)} \\
&= 4.38 - 0.40 \text{ (2.639)} \\
&= 3.324
\end{aligned}$$

By taking the antilogarithm of 3.324 we obtain the prediction

$$\hat{\text{COST}} = \text{antilog } (3.324) = \$27.80$$

Problems

6.1 Based on the estimated equation, determine the predicted percentage reduction in marginal cost each time production doubles.

6.2 Based on the estimated equation, determine the predicted marginal cost when cumulative production equals 1 million units. Use this result to provide an interpretation of the intercept in the equation.

6.3 Based on the estimated equation, what is the predicted marginal cost when cumulative production equals 15 million units?

6.4 Based on the estimated equation, what is the predicted marginal cost when cumulative production equals 16 million units?

The previous discussion suggests that substantive knowledge about a problem helps us anticipate the shape or functional form of a possible relationship. In addition, scatterplots of the data can be used to obtain information about the shape of the relationship, given sufficient data, and as long as there are no other relevant predictor variables. We may also be able to get clues about the correct shape by examining an estimated equation critically. For example, suppose we had estimated a linear equation for the relationship between marginal cost and cumulative production. Using the least-squares procedure directly on the data in Table 6.1, we obtain

```
> COEF

VARIABLE        B        STD. ERROR(B)      T

PROD       -.4503E+01     7.5400E-01     -5.972
CONSTANT    7.0467E+01    4.6785E+00     15.062

> SUMM

                MULTIPLE R    R-SQUARE

UNADJUSTED       0.9038        0.8168
ADJUSTED         0.8910        0.7939

STD. DEV. OF RESIDUALS = 6.84858
```

The regression output for this linear equation includes a respectable R^2 value of about 82 percent. There is also a computed t-value of -5.97, far beyond the critical value for standard type 1 error values. There is no indication from these results that there may be a systematic problem with this linear equation. However, we can compare the slope -4.5 with the data in Table 6.1 and see that it does not capture the pattern. For example, based on the data, as X increases from 1 to 2 the marginal cost declines by $19. And, when X increases from 9 to 10, marginal cost declines by only $1. Thus, for small values of X this slope coefficient underestimates the true effect, while for large values of X it overestimates the effect.

An examination of the intercept is also useful for the determination of the appropriateness of a linear equation. The value is 70.5, which suggests that the marginal unit cost is $70.50 when cumulative production is zero. In practice, it is often difficult to know this cost, but we do know that the actual marginal cost is $80 for 1 million units cumulative production (see Table 6.1). This comparison suggests that the intercept value is too low. The implications of this equation are particularly disturbing when the cumulative production continues to increase. As an example, suppose $X_i = 20$. The predicted marginal cost for this value equals $70.5 - 4.5 (20) = -19.5$, or a negative marginal unit cost. Given that the marginal unit cost has to be positive, this implication makes the equation unacceptable.

Problems

6.5 Predict the marginal cost when cumulative production is 15 million units, based on the equation CÔST = 70.5 − 4.5 PROD. Compare the prediction with the result from Problem 6.4, and suggest which prediction is most believable.

6.6 Predict the marginal cost when cumulative production is 16 million units, based on the equation CÔST = 70.5 − 4.5 PROD. Compare the prediction with the result from Problem 6.4, and suggest which prediction is most believable.

Although it is important to examine the implications of an estimated equation, one could argue that the equation is applicable only for a limited range of values. For example, one might suggest that the result is only valid for the range $1 \leq X_i \leq 10$. As we saw, however, even in that range the linear equation does not represent the observed data in Figure 6.3 well, and it does not enhance our understanding. In addition, we want to make predictions of marginal cost for future periods, when cumulative production exceeds the range observed in the sample. Thus, a disclaimer about limited applicability defeats the purpose of the study (predicting marginal cost for future cumulative production values), and even for the range of values observed in the sample, the linear relationship between the original variables is not a valid description.

Residual Plot

We can also discover a systematic problem with this linear equation by computing the residuals and comparing them to the corresponding values for the predictor variable. Shown below are the observed values for the criterion variable (cost) in the sample, the corresponding fitted values based on the linear model ($\hat{Y} = 70.5 − 4.5X$), the residual values, and the corresponding observed values for the predictor variable (production).

Y_i Observed value for criterion variable	$\hat{Y}_i$ Fitted value for criterion variable	$Y_i − \hat{Y}_i$ Residual value	X_i Observed value for predictor variable
80	66.0	14.0	1
61	61.5	−0.5	2
52	57.0	−5.0	3
46	52.0	−6.5	4
42	48.0	−6.0	5
39	43.4	−4.4	6
37	38.9	−1.9	7
35	34.4	0.6	8
33	29.9	3.1	9
32	25.4	6.6	10

If the proper shape of the relationship is indeed linear, then the residuals should not be systematically related to the values of the predictor variable. That is, at any given value of X the residual value should be just as likely to be positive as negative. Instead, we see that for $2 \leq X_i \leq 7$, all residuals are negative. The problem is quite evident in Figure 6.5, where the standardized residuals are plotted against the standardized values for cumulative production. The pattern suggests something like a U or V shape. We discuss in more detail the use of such residual plots for diagnostic purposes in Chapter 7.

Figure 6.5
Scatterplot of ten standardized values of 'residuals' vs. 'production'

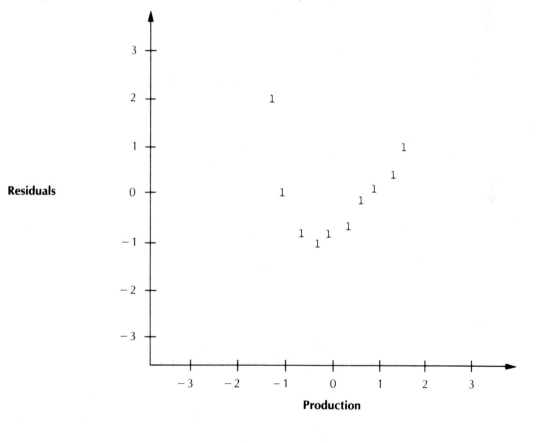

	MEAN	STD. DEV.
VERT. VAR.	0	6.84858
HORIZ. VAR.	5.5	3.02765
SAMPLE SIZE = 10		

Salary of Professional Basketball Players

In multiple regression applications, it is often more difficult to evaluate the soundness of a given functional form for a predictor variable. In addition, the use of a transformation for one predictor variable may have implications for other predictor variables. We illustrate some additional aspects in a study of the first-year salary for professional basketball players drafted in the first round. Specifically, this salary is assumed to be a function of three predictor variables: the player's position in the first round of the draft, whether the player was represented by a prominent lawyer, and whether an alternative professional league was in existence. The sample consists of 75 observations on players drafted by the National Basketball Association in the first round during a four-year period. Variable definitions and a matrix with all simple correlation coefficients are shown below.

```
                    VARIABLE DEFINITIONS

    MONEY — FIRST-YEAR SALARY IN THOUSANDS OF CURRENT DOLLARS

    DRAFT — POSITION IN FIRST ROUND OF THE DRAFT

            (1 = FIRST PLAYER SELECTED; 2 = SECOND; ETC.)

    PROM  — DUMMY VARIABLE:  1 IF PLAYER WAS REPRESENTED BY A
                                PROMINENT LAWYER

                             0 OTHERWISE

    ABA   — DUMMY VARIABLE:  1 IF PLAYER WAS SELECTED WHEN ABA WAS
                                IN EXISTENCE AS COMPETING LEAGUE

                             0 OTHERWISE


            CORRELATION MATRIX

            MONEY   DRAFT   PROM    ABA

    MONEY    1.00
    DRAFT   -0.61    1.00
    PROM     0.48   -0.17   1.00
    ABA      0.33   -0.06   0.05   1.00
```

The entries in the matrix, containing all simple correlation coefficients, suggest the following. Without holding the other predictor variables constant, the earlier a player is selected in the draft, the higher the salary ($r = -0.61$). Players represented by a prominent lawyer tend to have a higher starting salary ($r = 0.48$), and the starting salaries are higher when a competing league exists ($r = 0.33$). All simple correlations between the predictor variables are close to zero. Nevertheless, the interpretations based on the signs are as follows. Players drafted early are more likely to be represented by a prominent lawyer than players drafted later ($r = -0.17$). During the years that the American Basketball Association (a competing league) was in existence, the average value of DRAFT is somewhat lower ($r = -0.06$) than the average value during the years the ABA was not in existence. (This may be the result of some missing values.) And players are somewhat more likely to be represented by a prominent lawyer when the ABA was in existence ($r = 0.05$).

For the multiple regression, we propose that the first-year salary is determined by the player's draft position in the first round and the two other predictor variables. For example, holding the other two variables constant, the earlier the player is drafted, the higher the salary. Similarly, representation by a prominent lawyer is expected to increase the starting salary.[2] And, in years when a competing basketball league, the American Basketball Association, was in existence, salaries are expected to be higher. A linear multiple regression analysis confirms these expectations:

$$\text{MO\hat{N}EY} = 157 - 6.6 \text{ DRAFT} + 95 \text{ PROM} + 45 \text{ ABA}$$
$$(0.9) \qquad\qquad (18) \qquad\quad (13) \leftarrow \text{estimated standard errors}$$

For example, as the draft position increases by one (i.e., later selection) the starting salary tends to decline by $6,600, with a standard error of $900 (holding PROM and ABA constant). Similarly, the estimated effect of being represented by a prominent lawyer is $95,000, while the existence of the ABA league tends to add $45,000 to the starting salary.

On closer inspection, this equation has several problems. The intercept equals 157, which together with the coefficient of -6.6 for DRAFT implies that the *first* player drafted, *not* represented by a prominent lawyer, while the ABA was *not* in existence, could expect a starting salary of about $150,000. The same individual is predicted to have a first-year salary of about $290,000 with representation by a prominent lawyer during the existence of the ABA. This is the highest possible predicted first-year salary. Actually the highest salary in the sample is $500,000. Thus, the model appears to provide a substantial understatement of the observed salaries.

The linear equation also implies that the difference between two subsequent draft positions is always predicted to be $6,600, regardless of whether we compare the first and second players selected, or the eleventh and twelfth players chosen. In reality, it is likely that the difference between players selected early is much greater than the difference between players selected late in the draft. It also seems implausible that the benefit of representation by the prominent lawyer (or the existence of the ABA) is the same for all players. Instead, it seems more reasonable that the expected value of PROM is greater for players selected early than for players selected late in the first round. An examination of the implications from these results should convince us that this model is not satisfactory.

We may also discover the inappropriateness of a linear relationship between MONEY and DRAFT by examination of a scatterplot. Such a plot is shown in Figure 6.6. Strictly speaking, this plot is not a perfect basis for determining whether a linear relationship is acceptable, because there are other relevant predictor variables. However, neither of the two indicator variables is highly correlated with DRAFT (see the correlation matrix), and for that reason this plot can serve as an appropriate basis. The scatterplot in Figure 6.6 does suggest a nonlinear relationship between MONEY and DRAFT. This plot also suggests that the slope is negative and approaches zero as the value for DRAFT increases. Thus, a logarithmic transformation of the two variables may be used. We show below the definition of the transformed variables, the matrix of all simple

Figure 6.6
Scatterplot of 75 standardized values of 'money' vs. 'draft'

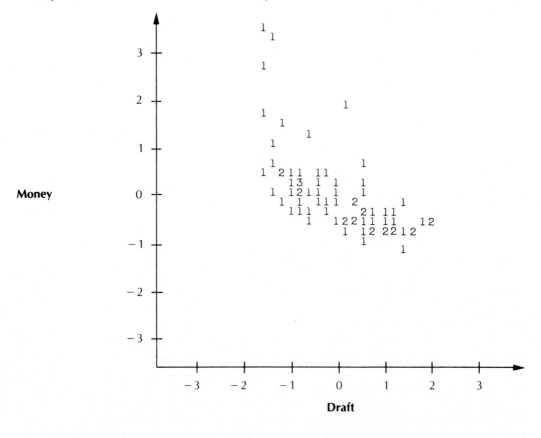

	MEAN	STD. DEV.
VERT. VAR.	115.133	71.8821
HORIZ. VAR.	10.1067	5.71526
SAMPLE SIZE = 75		

correlations, and multiple regression analysis results (with the transformed variables). Note that there is no need to transform the indicator variables.

VARIABLE DEFINITIONS

LMONEY — NATURAL LOGARITHM OF FIRST-YEAR SALARY

LDRAFT — NATURAL LOGARITHM OF POSITION IN FIRST ROUND

CORRELATION MATRIX

	LMONEY	LDRAFT	PROM	ABA
LMONEY	1.00			
LDRAFT	−0.76	1.00		
PROM	0.41	−0.18	1.00	
ABA	0.32	−0.04	0.05	1.00

EQUATION

$$\text{LMONEY} = 5.37 - 0.42\ \text{LDRAFT} + 0.44\ \text{PROM} + 0.32\ \text{ABA}$$

$$(0.04) \qquad\qquad (0.11) \qquad\qquad (0.09) \leftarrow \text{estimated standard errors}$$

$$R^2 = 0.73 \qquad R^{-2} = 0.72$$

To see if this multiple regression has reasonable implications, we examine again several aspects. The intercept is 5.37, suggesting that for a player drafted first in the first round (LDRAFT = 0 when DRAFT = 1, since ln(1) = 0), not represented by a prominent lawyer, when the ABA was not in existence, the predicted value of LMONEY is 5.37. Taking the antilogarithm, we obtain 215, or a starting salary of $215,000. And, if this player is represented by the lawyer, we obtain a predicted value of 5.81 for LMONEY, or $333,000. Thus the predicted value of PROM for a player drafted first is $118,000 in first-year salary. If we also add the expected value for the existence of the ABA, we obtain 6.13 for LMONEY, or a predicted starting salary of $460,000. This predicted value is very close to the maximum salary observed in the sample.

We also discover several appealing characteristics of the estimated equation with the transformed variables. A nonlinear relationship between MONEY and DRAFT appears to have been approximately accommodated, as we see shortly in Figure 6.7. At the same time, the transformations have also resulted in interaction effects for the original variables. For example, the expected effect of being represented by the prominent lawyer now depends on the position in the draft. To see this we compare two situations: (1) a player drafted first and (2) a player drafted tenth. For both cases, we assume the ABA is not in existence. The predictions are shown as follows.

1. A PLAYER DRAFTED FIRST (ASSUME ABA = 0)

$$\begin{array}{llll} \text{WITH PROM} & \text{LMONEY} = 5.81 & \text{MONEY} = 333 \\ \text{WITHOUT PROM} & \text{LMONEY} = 5.37 & \text{MONEY} = \underline{215} \\ & & 118 \end{array}$$

2. A PLAYER DRAFTED TENTH (ASSUME ABA = 0)

$$\begin{array}{llll} \text{WITH PROM} & \text{LMONEY} = 4.84 & \text{MONEY} = 126 \\ \text{WITHOUT PROM} & \text{LMONEY} = 4.40 & \text{MONEY} = 81 \\ & & 45 \end{array}$$

A comparison of the two cases shows that the predicted value of being represented by the lawyer is $118,000 for the player selected first, but only $45,000 for the player selected tenth. This decrease in expected value as the player is selected later is quite sensible. A similar interaction effect can be found for the ABA indicator variable.

Problems

6.7 Determine the predicted increase in first-year salary when the ABA is in existence, compared to when it is not, for a player drafted first (assume PROM = 0). Also determine the predicted increase when the ABA is in existence for a player drafted tenth (again assume PROM =0).

6.8 Determine the predicted increase in first-year salary for a player represented by a prominent lawyer, compared to when he is not, when the ABA is in existence (assume DRAFT = 1). Also determine the predicted increase when PROM = 1, for the case when ABA = 0 (again assume DRAFT = 1). Is the nature of the interaction effect reasonable?

Finally, we show a scatterplot of the standardized values for LMONEY and LDRAFT in Figure 6.7. The plot confirms that a linear relationship between these transformed variables is reasonable.

Innovation Diffusion

A different type of nonlinearity is associated with a study of the diffusion of an innovation. For example, we may be interested in a function that shows how the use of an innovation depends on the length of time the innovation has been available. Consider the case of a telephone company interested in setting sales targets for each of the

Figure 6.7
Scatterplot of 75 standardized values of 'LMONEY' vs. 'LDRAFT'

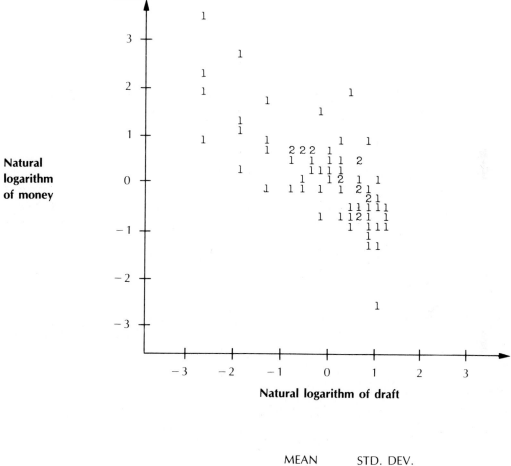

Natural logarithm of draft

	MEAN	STD. DEV.
VERT. VAR.	4.61788	.479585
HORIZ. VAR.	2.07352	.801815
SAMPLE SIZE = 75		

company's sales territories. The criterion variable of interest is the percentage of residence-telephone subscribers in the territory with touch-tone (as opposed to rotary-dial) equipment. Predictor variables might include the percentage of subscribers owning their residence, average family income, number of children, and so forth. In addition, the number of months touch-tone service has been available in a given sales territory is a predictor variable. To simplify the analysis we use only the last predictor variable for an illustration. Descriptive statistics are shown below:

```
VARIABLE         MEAN          STD. DEV.

TTP        42.6 (PERCENT)      13.5
TTAVL      27.7 (MONTHS)        8.3
```

where   TTP   IS TOUCH–TONE PENETRATION

  TTAVL   IS LENGTH OF AVAILABILITY OF TOUCH–TONE SERVICE IN
  MONTHS

In the sample, we see that there is considerable variation in the criterion variable. On the average, the percentage of households with touch-tone service is almost 43 percent, and the standard deviation is 13.5 percent. Also, the average number of months touch-tone service has been available is almost 28 months, but some territories have had it longer than that and some shorter. We want to learn something about how touch-tone penetration depends on the length of time the service has been available from a cross-sectional regression analysis (using data for several sales territories).

A linear regression analysis provides the following result:

$$\hat{TTP} = 21 + 1 \text{ TTAVL}$$

The intercept suggests that the penetration of touch-tone service is predicted to be 21 percent when the service is not available (TTAVL = 0). Clearly, this is implausible. Also, after, say, 100 months of availability, the prediction is that 121 percent of the households have the touch-tone service. Again, a study of the implications makes it obvious that a linear equation is not reasonable.

Having rejected a linear relationship between TTP and TTAVL, we have to identify an alternative shape for the possible relationship. Many studies of the diffusion of an innovation have identified an *S*-shaped curve, as shown in Figure 6.8. The proposed relationship suggests that as TTAVL increases, TTP tends to increase, but at different rates.

When the service has been available for only a short time (TTAVL is small), increases in TTAVL have a minimal effect on TTP. Then the slope accelerates, and it reaches its maximum value at approximately TTAVL*. After that point the effect of further increases in TTAVL declines, and it approaches zero when TTAVL becomes very large.

An appropriate model for this problem is

$$Y_i = e^{\alpha + \beta \frac{1}{X_i} + u_i}$$

where Y_i is TTP in territory i, and
  X_i is TTAVL in territory i.

Alternatively, we can substitute values for X and calculate Y, using plausible values for α

Figure 6.8
Diffusion of an innovation

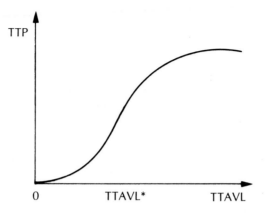

and β (see below). After taking natural logarithms of both sides we have

$$\ln Y_i = \alpha + \beta \frac{1}{X_i} + u_i$$

Thus, we perform a linear regression analysis *after* transforming both variables as indicated. To show that this is an appropriate functional form for the relationship illustrated in Figure 6.8, we can take derivatives of Y with respect to X. The first derivative (the slope) must be positive for all values of X. This can easily be shown to be true for $\beta < 0$. The second derivative shows the change in the slope, and by setting it equal to zero we obtain the point of inflection. (The point after which the change in the slope is negative, or the value for X where the first derivative is at its maximum.) This point can be shown to be $X = -\beta/2$. We can also substitute various values for X and determine the respective values for Y using plausible numbers for α and β (see Table 6.2).

Table 6.2
A diffusion example (using $\alpha = 4.55$ and $\beta = -20$)

X	$1/X$	$(\ln \hat{Y})$	$\hat{Y}$
1	1	−15.45	0
2	0.5	− 5.45	0
3	0.33	− 2.12	0.12
4	0.25	− 0.45	0.64
5	0.2	0.55	1.7
10	0.1	2.55	13
20	0.05	3.55	35
30	0.033	3.89	49
50	0.02	4.15	63
100	0.01	4.35	78
200	0.005	4.45	86

Application

Suppose we use the transformed variables and estimate the equation for touch-tone penetration. The result is

$$(\ln \hat{Y}_i) = 4.55 - 20 \, \frac{1}{X_i}$$

To see whether the estimated equation is sensible, we can check several aspects. For example, the predicted touch-tone penetration should never exceed 100 percent. If the service has been available for a long time (i.e., $X \to \infty$, and $1/X \to 0$) we obtain

$$(\ln \hat{Y}_i) = 4.55$$

Taking the antilog of 4.55, we obtain $\hat{Y} = 95$ (percent). Thus, the maximum possible penetration is predicted to be 95 percent. Also, as $X \to 0$, $1/X \to \infty$, and $(\ln \hat{Y}) \to -\infty$. As $(\ln \hat{Y}) \to -\infty$, $\hat{Y} \to 0$ (from above). Thus the predictions can never be below 0 percent.

To see more precisely the nonlinearity between penetration and availability based on the estimated equation, we can take specific values for X and show the corresponding transformed value $(1/X)$, the predicted value of $\ln Y$, and the predicted penetration, in Table 6.2.

Logistic Transformation

The logarithmic transformation of the criterion variable combined with the inverse of the predictor variable has been shown to allow for an *S*-shaped relationship between the original variables. We now consider a transformation, applied only to the criterion variable, that may be appropriate when that variable measures the probability of an event or the proportion of times an event occurs. The observed values for such a criterion variable are, of course, limited to the range [0, 1]. Nevertheless, an estimated equation may provide predicted (or fitted) values outside this range. To avoid such implausible predictions, we may prefer to use a transformation that guarantees all predictions to be within this range.

Another important aspect to consider when the criterion variable measures a probability is that the effect of a given predictor variable is not necessarily equal for all ranges of the criterion variable. For example, to achieve an increase from 0.05 to 0.06 may be a great deal more difficult than to achieve a change from 0.40 to 0.41. However, an increase from 0.94 to 0.95 may be just as difficult to achieve as it is to obtain an increase from 0.05 to 0.06. Such nonlinearity can be allowed for as well.

The logistic transformation of the criterion variable accommodates both aspects discussed above. The transformation is

$$P^* = \ln \frac{P}{1 - P}$$

where P is the observed probability

With this transformation we note the following. The original criterion variable has the constraint $0.0 \leq P \leq 1.0$, in terms of actual, observable values. It follows that the ratio $P/(1 - P)$ is constrained between 0 and ∞. Then the natural logarithm of this ratio is between $-\infty$ and $+\infty$. Now, working backwards, all *predictions* of the quantity $\ln P/(1 - P)$ vary between $-\infty$ and $+\infty$. Transforming back to $P/(1 - P)$ and P, we can only obtain predicted values for P between 0 and 1.

To demonstrate the nature of the transformation, we have used some values of P and show the corresponding values of $P/(1 - P)$ and $\ln P/(1 - P)$ in Table 6.3. The last column in this table shows that the difference between 0.40 and 0.41 after the logistic transformation equals 0.04, while the difference between 0.05 and 0.06 after the logistic transformation is 0.19. But the difference between 0.94 and 0.95 after the logistic transformation is again 0.19. Thus, all three differences are equal in terms of the difference in probabilities (0.01 in each case), but the difference is more dramatic, after the logistic transformation, for probability values close to zero. Similarly, the difference is more dramatic for probability values close to one.

The values shown in Table 6.3 have also been used to construct the graph in Figure 6.9. This graph makes it clear that predicted values of P are constrained to be

Table 6.3
Illustrative values for the logistic transformation

P	$\dfrac{P}{1-P}$	$\ln\left(\dfrac{P}{1-P}\right)$	Difference[1] between ln value and previous ln value
0.01	0.0101	-4.61	
0.02	0.0204	-3.91	0.70
0.05	0.053	-2.94	
0.06	0.0638	-2.75	0.19
0.10	0.1111	-2.20	
0.11	0.1236	-2.09	0.11
0.20	0.25	-1.39	
0.21	0.2658	-1.32	0.07
0.40	0.6667	-0.40	
0.41	0.6949	-0.36	0.04
0.50	1.0	0	
0.59	1.439	0.36	
0.60	1.5	0.40	0.04
0.79	3.762	1.32	
0.80	4.0	1.39	0.07
0.89	8.091	2.09	
0.90	9.0	2.20	0.11
0.94	15.667	2.75	
0.95	19.0	2.94	0.19
0.98	49.0	3.91	
0.99	99.0	4.61	0.70

[1] When the difference between successive P-values equals 0.01.

Figure 6.9
Relationship between P and $\ln P/(1 - P)$

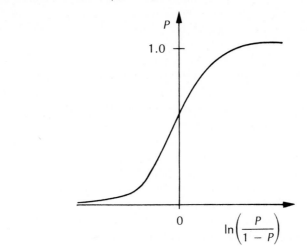

in the range $[0, 1]$, and that gains in P are easier in the middle of the range than at either end. Note that we only show the relationship between the original variable, P, and the transformed variable, $\ln P/(1 - P)$, in Figure 6.9. Whatever predictor variable(s) we use should be linearly related to $\ln P/(1 - P)$.

An Inverted-*U* Relationship

For certain problems it is important to allow the slope to change *signs*. We show a simple example of such a relationship in Figure 6.10. The graph in Figure 6.10 shows that for small values of X there is a positive relationship between Y and X. However, for large

Figure 6.10
An inverted-*U* relationship

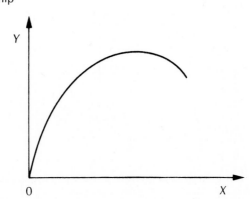

values of X the relationship is negative. And the maximum possible value of Y, based on X, exists at the point where the slope is zero (or the point where the sign of the slope changes from positive to negative). This type of relationship can be accommodated by adding the squared values of X as a separate predictor variable to the model. Assuming that X is the only relevant predictor variable, we have

$$Y_i = \alpha + \beta_1 X_i + \beta_2 X_i^2 + u_i$$

If $\beta_1 > 0$ and $\beta_2 < 0$, then for $\beta_1 > |\beta_2|$, the initially positive relationship between Y and X changes into a negative relationship when

$$X_i = -\frac{\beta_1}{2\beta_2}$$

Application

Suppose we are charged with hiring employees for a company's new division, which has been in operation for only six months. One of our tasks is to decide how much education is desirable for employees who assemble a small electronic device. We believe that the best employees are high-school graduates. That is, we expect individuals who failed to complete high school to be less productive than high-school graduates. We also believe that employees with formal education beyond high school are over-qualified and will not perform as well as high-school graduates.

To investigate this hypothesis, we have taken a sample of 21 employees whose formal education varies from six to fourteen years. They have all worked for the new division an equal amount of time. We assume, for the purpose of this illustration, that there are no other relevant predictor variables. For each employee we have a measure of the average number of units produced per day. The model is

$$Y_i = \alpha + \beta_1 X_i + \beta_2 X_i^2 + u_i$$

where Y_i is the average number of units produced per day by employee i
(OUTPUT),

X_i is the number of years of education for employee i (EDUC), and

X_i^2 is the squared number of years of education for employee i (EDUCSQ).

Results are shown below.

```
> COEF
```

VARIABLE	B	STD.ERROR(B)	T
EDUC	4.7029E+01	4.2054E+00	11.183
EDUCSQ	−.1937E+01	2.0445E−01	−9.476
CONSTANT	3.8651E+01	2.0582E+01	1.878

Based on the computer output, the estimated equation is

$$\hat{Y}_i = 38.7 + 47.0X_i - 1.9X_i^2$$

To estimate the total effect of education on daily employee output, we combine the estimated effects for X and X^2. These effects are shown below for selected levels of education.

Education in years	X_i	X_i^2	Estimated education effect
6	6	36	214
10	10	100	280
11	11	121	287
12	12	144	290
13	13	169	290
14	14	196	286
15	15	225	277
19	19	361	207

(12 and 13 marked as }maximum)

From the estimated effects we see that increases in education tend to increase productivity until the amount of education reaches 12 or 13 years. Beyond that point, further education decreases productivity. The change from a positive to a negative relationship happens when the level of formal education reaches $-\dfrac{47}{(2)(-1.9)} = 12.4$ years.

Problem

6.9 For the estimated relationship between productivity and level of education, predict the number of units produced per day by an employee with nine years of education.

Indicator Variables for Nonlinear Effects

Sometimes we consider using indicator variables to allow for nonlinear effects. This is especially appropriate if there is considerable uncertainty about the nature of the nonlinearity, and if the number of distinct categories is small. For example, suppose we want to study the relationship between income and consumption of a particular item. Assume that a survey of many families has been taken to obtain measures of consumption and information on annual household income. We expect income to have a positive relationship with consumption of the item, but not in a linear form. However, we have great difficulty imagining an appropriate transformation for any nonlinearity that may exist. The problem is compounded by the fact that households only checked one of the following five categories to supply income information:

Category	Household income (dollars)
1	≤9,999
2	10,000 − 19,999
3	20,000 − 29,999
4	30,000 − 49,999
5	≥50,000

If we want to use income as a predictor variable, we have to substitute typical income values for category membership. For example, for a family checking income category 1, we could substitute the middle value of the category, or $5,000. Of course it is especially difficult, without more information, to determine an appropriate value for category 5. For that reason, we may consider using the numbers for the categories to capture the different income categories and then to propose a particular relationship.

Given our lack of knowledge about the shape of the relationship of interest, we may prefer to use four indicator variables for income. For example, let

$D_{2i} = 1$ if household i checked category 2
    $= 0$ otherwise
$D_{3i} = 1$ if household i checked category 3
    $= 0$ otherwise
$D_{4i} = 1$ if household i checked category 4
    $= 0$ otherwise
$D_{5i} = 1$ if household i checked category 5
    $= 0$ otherwise

We can now construct two alternative models and compare them in terms of explanatory power. For the illustration, we assume a linear relationship between consumption and the income categories (1 through 5) which implicitly allows for some nonlinearity between consumption and income, and compare this model with one using four indicator variables.

The model with the four indicator variables is

$$\text{Model I} \quad \hat{Y}_i = a + b_2 D_{2i} + b_3 D_{3i} + b_4 D_{4i} + b_5 D_{5i}$$

Similarly, the model with a linear effect for the income categories is

$$\text{Model II} \quad \hat{Y}_i = a + b_1 X_i$$

Assuming that income is related to consumption, we now want to determine whether we should rely on model I or II. Model I has the advantage of allowing for maximum flexibility in the shape of the relationship. Model II incorporates the restriction that the difference in consumption between households in adjacent income categories is always b_1. That is, the difference between categories 1 and 2 equals the difference between categories 2 and 3, and so forth. On the other hand, model II has the advantage of using only one degree of freedom (one slope coefficient), whereas model I requires four degrees of freedom.

Intuitively, we can resolve this problem as follows. If model II provides the same explanatory power as model I, then we should prefer model II. Thus, if the two models have about the same R^2 values, we can conclude that the restriction of linearity between Y and the categories of X is appropriate. On the other hand, if the explanatory power of model I is substantially greater than that of model II, we may conclude that the flexibility of model I is preferable.

We can also examine the appropriateness of linearity by comparing the values of the coefficients b_2 through b_5 in model I. For example, b_2 provides an estimate of the difference in consumption between categories 1 and 2. Similarly, b_3 is an estimate of the difference in consumption between categories 1 and 3. Now, linearity implies that $(b_3 - b_2) = b_2$. And, for all categories, linearity implies that $(b_5 - b_4) = (b_4 - b_3) = (b_3 - b_2) = b_2$.

A Test

To determine which model should be preferred, we may use an incremental R^2 test. The null hypothesis is

H_o: $(\beta_5 - \beta_4) = (\beta_4 - \beta_3) = (\beta_3 - \beta_2) = \beta_2$

H_A: at least two of these terms are different from each other

The null hypothesis states that the relationship between the income categories and consumption is linear. The alternative hypothesis embodies the flexibility associated with the use of indicator variables. We know that the unadjusted R^2 for model I is at least as high as the unadjusted R^2 for model II, because model II is a restricted form of model I. The question is whether the additional explanatory power is greater than may be due to chance. The test statistic is

$$F = \frac{(R_I^2 - R_{II}^2)/(DF_{II} - DF_I)}{(1 - R_I^2)/DF_I}$$

where R_I^2 is the unadjusted R^2 for model I

R_{II}^2 is the unadjusted R^2 for model II

DF_I is the number of degrees of freedom left[3] in model I

DF_{II} is the number of degrees of freedom left in model II

If the computed F-value exceeds the tabulated F-value for $(DF_{II} - DF_I)$ and DF_I degrees of freedom at the desired level of significance, we reject the null hypothesis and choose model I over model II.

Problem

6.10 Show how the following model allows for nonlinearity between consumption and income: $\hat{Y}_i = a + bX_i$

where Y_i is consumption of household i, and

X_i is the number of the category the household checked regarding income.

Relative Gain as a Criterion Variable

For the purpose of identifying appropriate transformations, we have focused on the theoretical shape of a relationship between the criterion variable and a given predictor variable. Sometimes, however, a transformation of the criterion variable is required even though we do not necessarily anticipate nonlinearity. To illustrate this, we provide an example of the *measurement* of a criterion variable.

Suppose we are interested in a model that uses relative change in market share for a brand as the criterion variable.[4] Three market share observations are available for several brands, and we are interested in explaining differences in the (average) relative change in market share between the brands. We focus only on the criterion variable, since the identity of predictor variables is irrelevant. Consider the hypothetical data shown in Table 6.4.

In Table 6.4 we consider two brands, A and B. The brands end at the same market share level in period 3 as they started in period 1. Thus, from the first to the last period, neither brand gains nor loses. However, if we compute the period-to-period relative

Table 6.4
Average relative changes in market shares

t Time period	$MS_{A,t}$ Market share brand A	$MS_{B,t}$ Market share brand B	$\left(\dfrac{MS_{A,t} - MS_{A,t-1}}{MS_{A,t-1}}\right)$ Relative change in market share brand A	$\left(\dfrac{MS_{B,t} - MS_{B,t-1}}{MS_{B,t-1}}\right)$ Relative change in market share brand B
1	15	55	—	—
2	20	50	+33.3%	−9.1%
3	15	55	−25.0%	+10.0%
Average			+4.2%	+0.5%

changes (see the last two columns in Table 6.4), we obtain an average relative gain for brand A of 4.2%, and an average relative gain of 0.5% for brand B. (Of course, if we compute the relative gain from period 1 to 3 directly we observe no gain.) From this example, we conclude that it is inappropriate to compute an arithmetic mean for a variable that measures the relative gain. And if it is inappropriate to compute an arithmetic mean for the criterion variable, it is inappropriate to use least-squares regression analysis[5] without some transformation.

Instead of using

$$\frac{MS_t - MS_{t-1}}{MS_{t-1}}$$

for a given brand, we can define

$$\frac{MS_t}{MS_{t-1}}$$

The latter ratio is never negative, so we can take the *geometric* mean. The geometric mean is defined as

$$\sqrt[n]{\prod_{i=1}^{n} Y_i}$$

where Y_i is the variable of interest

n is the number of observations

Π is the multiplication symbol

$\sqrt[n]{}$ is the nth root

For the example involving relative market shares, we have

$$\sqrt{\left(\frac{MS_{t+2}}{MS_{t+1}}\right)\left(\frac{MS_{t+1}}{MS_t}\right)} = \sqrt{\frac{MS_{t+2}}{MS_t}}$$

Thus, the geometric mean turns out to be based on the ratio of the ending value relative to the starting value. For the data in Table 6.4, the geometric mean is zero for both brands A and B.

Given the multiplicative nature of the geometric mean, it is easy to see that a logarithmic transformation makes the formula additive. That is

$$\ln \sqrt[n]{\prod_{i=1}^{n} Y_i} = \frac{\sum_{i=1}^{n} \ln Y_i}{n}$$

It follows that when the criterion variable measures relative performance (e.g., the ratio of two values measured in successive periods), a logarithmic transformation of the criterion variable is indicated. Using the logarithmic values as the criterion variable avoids the bias observed for the original variable.

Problems

6.11 Consider two brands, *C* and *D*, with market shares in three successive periods of 25, 20, 25 for *C*, and 45, 50, 45 for *D*. Compute the average relative change in market share for the two brands. Compare these arithmetic means with the geometric means for the ratio of market shares in successive periods.

6.12 In finance, it is common for researchers to study the return on investment in a given security as a function of the return on investment in a portfolio of securities. Let the security represent common stock for a given firm. Specify the criterion variable and predictor variable for the relationship of interest. Indicate what transformation should be considered, and provide definitions of the variables to be transformed.

SUMMARY

In Chapters 4 and 5 we have considered the possible consequences of omitting relevant predictor variables. For proper understanding of the effects of any predictor variable on the criterion variable, all relevant predictor variables should be included in the model. For an application of regression analysis, we recommend therefore that all relevant predictor variables be identified and included. However, once the set of predictor variables is defined, it is just as important to consider the shape of the effect of the predictor variable on the criterion variable for each predictor variable separately. In this chapter we have seen examples of misleading results obtained when the functional form is not properly specified. For the identification of the predictor variables and the specification of functional forms, we depend on our substantive knowledge of the problem we are studying.

Once these steps have been taken, it is important to identify appropriate transformations for each of the predictor variables that are thought to have a nonlinear relationship with the criterion variable. It may be necessary to transform the values of the criterion variable as well. After data have been obtained, the multiple regression equation can be estimated. As we argued in Chapter 5, we then test the overall equation for statistical significance. In the context of testing a hypothesis about a particular slope parameter, we have to evaluate the adequacy of the functional forms assumed and the transformations used, if any. We can do this by evaluating the implications of the estimated equation (e.g., implausible predictions or implications), by examining the data, and by checking and testing the residuals for systematic variation (see also Chapter 7).

As we have indicated in this chapter, scatterplots of the criterion variable against a predictor variable, examined to learn about the nature of the effect, may be misleading. Just as an omitted variable tends to cause problems for estimating the effects of predictor variables included in an equation, such an omission also causes problems for the discovery of the proper functional form. Not only must the effects of all relevant predictor variables be adjusted for, those effects must also be accommodated in the proper manner (e.g., nonlinear). In addition, the plots may be incomplete if a predictor variable's range of variation in the sample is limited. That is, the smaller the range, the more difficult it is to learn the nature of the relationship, using sample data. For these reasons it is crucial that we give careful consideration to the specification of the most reasonable functional forms for each predictor variable *before* we use the data.

ADDITIONAL PROBLEMS

6.13 A female faculty member, charging her university with sex discrimination in salary levels, conducted a regression analysis to show that female faculty members tended to be underpaid at the university, relative to male faculty members. The following model was used[6]:

$$Y = \alpha + \beta_1 X_1 + \beta_2 X_2 + u$$

where Y is annual salary in dollars,

X_1 is seniority in years, and

X_2 is an indicator variable (1 for women, 0 for men).

The empirical results included a statistically significant estimate equal to $-\$443$ of β_2 for a given year.

a. What is the interpretation of the parameter estimate, assuming that the model is correctly specified and that the assumptions are satisfied?
b. Suppose that (1) annual increases in salary tend to be calculated as a percentage of current salary, (2) the fraction of female faculty members at the university has grown steadily, and (3) a disproportionate part of the faculty has little seniority. Suggest a different model, with the same variables, that incorporates this knowledge, and provide a different explanation (relative to Problem 6.13*a*) for the result obtained with the linear model.

6.14 Specify an appropriate functional form, and show the transformation to be used for estimating the relationship between a firm's dividends and earnings per share for each case below.

a. The firm has a policy of using an approximately constant payout ratio (i.e., dividends are approximately equal to x per cent of profit).

b. The firm has a policy of increasing dividends by y percent of their previous level each time earnings double (e.g., if earnings increase from \$1 to \$2 per share, dividends increase by y percent compared to dividends for \$1 of earning).

6.15 Suppose that the following model is used to estimate the effects of the size of a house and its location on the sales price. Specifically, the estimated equation is

$$\ln (\text{Price}) = a + b_1 \ln (\text{Square feet}) + b_2 (\text{Location})$$

where   Price is the sales price in dollars,

Square feet is a measure of the floor space, and

Location is an indicator variable (1 = suburb, 0 = downtown).

In the study, $b_2 = 0.50$.

a. Assume that the coefficient for location is statistically significant, and that there is no evidence that other assumptions required for statistical inference are violated. What is the price difference between two otherwise equivalent homes, one of which is located in a suburb, the other downtown?

b. If $0 < b_1 < 1$, show graphically how location changes the relation between price and size of a home.

APPENDIX 6.1
COMPUTATION OF PERCENTAGE DECLINE

Question

By what percentage does the marginal cost decline when cumulative production doubles, if the slope of the transformed variable is -0.50?

Answer

The estimated equation is $\hat{Y}_i = aX_i^b$

Let

$$X_1 = 1 \text{ (first unit), and } X_2 = 2$$

Then

$$\hat{Y}_1 = aX_1^b = a, \text{ since } X_1 = 1$$

and

$$\hat{Y}_2 = aX_2^b = (\hat{Y}_1)(2)^b$$

Assume $\hat{Y}_1$ (cost of the first unit) equals \$100. Let Y_2 equal the cost of the second unit. Then

$$\hat{Y}_2 = (\hat{Y}_1)(2)^b \quad \text{or}$$

$$\hat{Y}_2 = (\$100)(2)^{-0.5}, \text{ since } b = -0.5$$

$$\ln \hat{Y}_2 = \ln \$100 - 0.50 \ln 2$$
$$= 4.605 - 0.50(0.693)$$
$$= 4.258$$

$$\hat{Y}_2 = \$70$$

Based on the calculations, the marginal cost is approximately 70 percent of what it used to be each time the cumulative production doubles.

APPENDIX 6.2
TAKING DERIVATIVES TO EXAMINE NONLINEARITY

The model is

$$Y_i = e^{\alpha}X_i^{\beta}e^{u_i}$$

and the estimated equation is

$$\hat{Y}_i = e^a X_i^b$$

Taking the first derivative of $\hat{Y}$ with respect to X,

$$\frac{\partial \hat{Y}_i}{\partial X_i} = e^a b X_i^{b-1}$$

The expression for the first derivative provides a measure of the slope. Since the predictor variable is included in this expression, it is clear that the slope depends on the value of X. The nature of this change depends on the value for b, as suggested below (for $X_i > 0$).

Value of b	Slope
$b > 1$	positive and increasing as X increases
$0 < b < 1$	positive, but decreasing as X increases
$b < 0$	negative, but approaching zero as X increases

ENDNOTES

1. The standardized values are obtained, for a given variable, by subtracting the sample mean and dividing the difference by the standard deviation for the variable in the sample.

2. We recognize that the selection of a prominent lawyer may also depend on salary expectations. This would imply problems with the assumed causal direction.

3. The degrees of freedom left refers to the number of observations in the sample minus the total number of parameters estimated. For model I, five parameters are estimated, compared to two for model II.

4. Buzzell, Robert D., and Frederik D. Wiersema, "Modeling Changes in Market Share: A Cross-Sectional Analysis," *Strategic Management Journal*, 2 (1981): 27–42.

5. The sample mean is the value that minimizes the sum of squared deviations between the individual values and that value. In a model with only an intercept (no predictor variables), the intercept is the arithmetic mean.

6. Barnett, Arnold, "Misapplication Reviews: The Linear Model and Some of Its Friends," *Interfaces*, 13 (February 1983): 61–65; and "Errata: The Linear Model and Some of Its Friends," *Interfaces*, 13 (June 1983): 90.

REFERENCES

Barnett, Arnold. "Misapplication Reviews: The Linear Model and Some of its Friends," *Interfaces*, 13 (February 1983): 61–65.

———."Errata: The Linear Model and Some of its Friends," *Interfaces*, 13 (June 1983): 90.

Buzzell, Robert D., and Frederik D. Wiersema. "Modeling Changes in Market Share: A Cross-Sectional Analysis." *Strategic Management Journal* 2 (1981): 27–42.

7

Diagnostics in Regression Analysis

In Chapter 3 we stated four assumptions about the error term in the model. All statistical inference is conditioned on the validity of these assumptions. In this chapter we describe the role of each assumption and discuss possible reasons for the violation of one or more assumptions in an empirical setting. We propose procedures that may be used to identify violations of these assumptions and consider the implications of such violations. Once we know that an assumption is violated and know the reason for this violation, it may be possible to make appropriate corrections in the model formulation. If appropriate corrections are available (e.g., transformation of variables), the standard assumptions must apply to the error term of the reformulated model.

We assume the general linear model

$$Y_i = \alpha + \beta_1 X_{1i} + \beta_2 X_{2i} + \cdots + \beta_m X_{mi} + u_i$$

The assumptions about the error term are

ASSUMPTION 1: $E(u_i) = 0$ for all i
ASSUMPTION 2: $\text{Var}(u_i) = \sigma^2$ for all i
ASSUMPTION 3: $\text{Cov}(u_i u_j) = 0$ for all $i \neq j$
ASSUMPTION 4: u_i is normally distributed for all i

Of these assumptions, by far the most critical one is the first. As we have argued in Chapter 3 as well as in subsequent chapters, this assumption is necessary for the estimates to be unbiased. For example, with the estimated equation

$$\hat{Y}_i = a + b_1 X_{1i} + b_2 X_{2i} + \cdots + b_m X_{mi}$$

we can only claim that $E(b_1) = \beta_1$ if $E(u_i) = 0$. The same requirement holds for the other estimates to be unbiased.

The requirement that estimates be unbiased is relevant both for statistical inference and for managerial decisions. With regard to statistical inference, consider the confidence interval around the estimated value b_1:

$$b_1 \pm t s_{b_1}$$

The interval is centered around the particular value for b_1 obtained from the regression analysis. If our estimate is biased, then this interval is meaningless. The true but unknown value for β_1 is more likely to be outside this interval, if b_1 is biased, than is indicated by the confidence level used to construct the interval. Managerial decisions are also compromised. If b_1 is a biased estimate, decisions based on the value of b_1 are wrong.

Reasons for $E(u_i) \neq 0$

Although there are many possible reasons for a violation of the first assumption, we have focused on the role of omitted predictor variables (Chapters 4 and 5) and the assumed functional form (Chapter 6). If a relevant predictor variable is omitted from the model, the parameter estimates tend to be biased. The magnitude of the bias depends on the nature and magnitude of the influence of the omitted variable on the criterion variable, and the magnitude of its correlation with the included predictor variables. We have also discussed in some detail the bias resulting from an incorrect functional form for the relationship between the criterion variable and a predictor variable.

Unfortunately, the data are only of limited use for the detection of these problems. Consequently, it is very important for the user of regression analysis to identify all relevant predictor variables, and to specify appropriate functional forms, before starting the empirical work. The following example suggests why sometimes the data are of limited use. We consider first the problem of the functional form. Suppose that the *true* relationship between a criterion variable and three predictor variables is

$$Y_i = \alpha + \beta_1 \ln X_{1i} + \beta_2 \frac{1}{X_{2i}} + \beta_3 X_{3i} + \beta_4 X_{3i}^2 + u_i$$

If instead we use

$$Y_i = \alpha + \beta_1 X_{1i} + \beta_2 X_{2i} + \beta_3 X_{3i} + u_i$$

then $E(u_i)$ for the second equation cannot be zero for all observations. To detect functional form problems we can plot the residuals of the second equation against each of the three predictor variables separately. These plots can be used to see if, after having allowed for a linear effect for each of the three predictor variables, there is evidence of additional systematic variation remaining in the residuals. However, given that the assumed functional form is wrong for *each* predictor variable in the second equation, these plots may be difficult to interpret. Also, with limited variability in the sample for a predictor variable, systematic problems become even more difficult to identify. Even if it is possible to identify problems with the assumed functional form, it is extremely difficult to infer from these plots what an appropriate functional form is.

   If there are omitted variables in addition to the functional form problems, interpretation of these residual plots may be close to impossible. For example, if we omit X_{3i} from the second equation, we have

$$Y_i = \alpha + \beta_1 X_{1i} + \beta_2 X_{2i} + u_i$$

The residuals from this equation will contain the influence of X_3 that is not picked up by the linear effects for X_1 and X_2, and the nonlinear part of the influence of X_1 and X_2. Thus, in this case, there is systematic variation in the residuals, with the amount and nature of this variation dependent upon the correlation between X_3 and the included predictors, and upon the extent of nonlinearity in the true influence of the included variables.

   If X_3 is highly correlated with the included predictors, there cannot be a great deal of systematic variation due to this omitted variable left in the residuals. Thus, the higher the correlation between included and omitted predictors, the lower the amount of systematic residual variation. Yet it is precisely when this correlation is high that the bias in the parameter estimates for the included predictors is the greatest. Also, if we examine the plot of residuals against the two predictors, we may infer that a quadratic term for one of these variables should be added to the model. More precisely, if X_3^2 is a relevant but omitted predictor, and X_3 is correlated with X_2, the residual plot may give the appearance that X_2^2 should be added to the equation.

   The greater the number of relevant predictors omitted, and the greater the difference between the assumed (e.g., linear) and the correct (e.g., quadratic) functional forms, the more difficult it is to interpret the residual plots. Also, the smaller the range of variation in the sample for a given predictor variable, the more difficult it is to test the validity of the assumed functional form. These considerations suggest that it is extremely important to spend a great deal of time on the initial model specification. The closer the initial specification is to the "correct" specification, the more useful an analysis of the residuals will be.

Plots

The previous discussion suggests that we should spend a considerable amount of time thinking about (1) the relevant predictor variables and (2) the appropriate functional form for each predictor variable. This up-front thinking should minimize the mistakes made in the initial model specification. And the fewer the mistakes the more likely it is that residual plots will be informative. For example, if all the relevant predictor variables are included in the model, and only one functional form is incorrect, all residual plots should have an acceptable pattern, except for the plot of residuals against the predictor variable for which the functional form is incorrect. In that case this plot should provide a good basis for making a correction in the model.

Tests

Even without the possibility of omitted variables and of functional form problems, the interpretation of a plot is subjective. Thus, it is useful to employ tests of the model specification as well. Sometimes tests of model specification or functional form are simple to perform. For example, if we choose between a linear effect and a quadratic effect for a predictor variable, we only have to test the null hypothesis that the parameter associated with the squared term is zero. Or, if we want to consider the possibility that two predictor variables each have a quadratic effect, we can use an incremental F-test for the null hypothesis that both quadratic effects are zero (see Chapter 6). We now consider an example where the choice between two alternative model specifications is not so straightforward.

Example

We use the case of a ballpoint-pen manufacturer interested in determining the marginal effectiveness of certain marketing efforts.[1] The manufacturer wants to determine the effects of variations in the number of television advertising spots per month and the number of sales representatives on ballpoint-pen sales. The available data represent a sample of 40 sales territories. For the illustration we assume that ballpoint-pen sales depend only on the number of television advertising spots and the number of sales representatives. That is, we consider the model correctly specified except for uncertainty about the functional form. We now consider two alternative models:

$$\text{Model I} \quad \hat{Y}_i = a + b_1 X_{1i} + b_2 X_{2i}$$

$$\text{Model II} \quad \hat{Y}_i = a + b_1 \ln X_{1i} + b_2 \ln X_{2i}$$

where Y_i is monthly ballpoint-pen sales in territory i,

X_{1i} is number of television advertising spots per month in territory i, and

X_{2i} is number of sales representatives in territory i.

Results are shown below.

$$\text{Model I} \quad \hat{Y}_i = 69.3 + 14.2 X_{1i} + 37.5 X_{2i}$$

$$R^2 = 0.8739$$

$$\bar{R}^2 = 0.87$$

$$\text{Model II}\quad \hat{Y}_i = -212.5 + 158.5 \ln X_{1i} + 165.9 \ln X_{2i}$$

$$R^2 = 0.8563$$

$$\bar{R}^2 = 0.85$$

The choice between these two models is not a simple one. We assume there are no theoretical arguments to prefer one specification over another. One basis for choosing between the two models is the R^2 values. This is possible if the criterion variables are expressed identically in the two models, as is the case here. Also, given that both models have two predictor variables, we can use either the unadjusted or adjusted R^2 values. Based on this goodness-of-fit criterion we find that model I is superior. Thus, it appears to be more appropriate to express X_1 and X_2 linearly, than it is to use logarithmic transformations for X_1 and X_2. Of course, the superiority of model I would be indicated more strongly if its goodness-of-fit value is significantly higher than the one for model II. Unfortunately, one model cannot be expressed as a restricted form of the other, making it impossible to use the incremental R^2 test directly. One other difficulty is that neither model may be satisfactory, and a comparison of the R^2 values can only aid in discovering the *relative* superiority of one model over another.

A solution to this problem is to specify a supermodel that contains the union of all predictor variables. This supermodel may not be of interest in itself; it would only be used to determine the adequacy of each alternative model. One attractive feature of the supermodel is that both models I and II may be rejected. If this happens, we conclude that neither model provides an adequate representation of the relationship. The supermodel is

$$\text{Model III}\quad \hat{Y}_i = a + b_1 X_{1i} + b_2 X_{2i} + b_3 \ln X_{1i} + b_4 \ln X_{2i}$$

We can now perform an incremental R^2 test. Specifically, we test the null hypothesis

$$H_o: \beta_3 = \beta_4 = 0$$

based on a comparison of R^2 values for models I (the best of the two models) and model III.

The results for model III are

$$\hat{Y}_i = 18.8 - 4.44 X_{1i} + 112.8 X_{2i} + 201.6 \ln X_{1i} - 378.3 \ln X_{2i}$$

$$R^2 = 0.9005$$

$$\bar{R}^2 = 0.89$$

We note that two of the signs for the supermodel are theoretically incorrect. This is a result of the multicollinearity introduced by having both linear and logarithmic values for the same predictor variable in the same equation. However, our intent is only to determine the adequacy of model I using model III as a comparison. We are not interested in interpreting the slope coefficients in model III.

The test statistic is

$$F = \frac{(R^2_{\text{III}} - R^2_{\text{I}})/(DF_{\text{I}} - DF_{\text{III}})}{(1 - R^2_{\text{III}})/DF_{\text{III}}}$$

where R^2_{I} is the unadjusted R^2 for model I

R^2_{III} is the unadjusted R^2 for model II

DF_{I} is the degrees of freedom left for model I

DF_{III} is the degrees of freedom left for model III

Using the results reported above, we have

$$F = \frac{(0.9005 - 0.8739)/2}{0.0995/35} = 4.75$$

This computed F-value exceeds the critical value of 3.32 at the five-percent level of significance. Thus, we reject the null hypothesis and conclude that model I is an inadequate specification.[2] Based on this test, we conclude that both models I and II are inadequate. (If model I is rejected, model II will also be rejected, given its lower R^2 value.) We conclude that more initial thought is required to obtain an alternative specification.

Interpretation

The supermodel makes it possible to use the incremental R^2 test introduced in Chapter 5. If model I is in fact the correct model specification, we expect the addition of the logarithmic terms to increase R^2 by a negligible amount. If model I is incorrect and the logarithmic terms contain at least part of what is missing from model I, we have a chance of rejecting it. Given that we have rejected model I at the five-percent level, we have to conclude that the relationship cannot be captured with linear or logarithmic effects alone. Perhaps the true relationship should be estimated with an entirely different formulation. However, it is also possible that there is an omitted variable which is correlated with the logarithmic or the linear specification or both. Yet another possibility is that the model requires a linear specification for one predictor variable but a logarithmic one for the other. Clearly we cannot resolve this dilemma unless we explore the problem in more detail, especially in terms of other relevant predictor variables or additional difficulties with the model specification (for example, the assumed direction of causality). The test has indicated merely that neither model I nor model II is a satisfactory representation of the relationships of interest.

This example is useful in that it illustrates a way in which poorly specified models can be tested and rejected. If the attempt to reject a model is not used, we are likely to make conclusions based on misleading (wrong, biased) results. In the example, the analysis suggests that neither model specification is appropriate. On the other hand, no guidance is available from the data as to what formulation to use instead. Substantive knowledge about the problem is required to identify relevant predictor variables and to specify functional forms, as well as to determine appropriate measures for the variables.

Reasons for $\text{Var}(u_i) \neq \sigma^2$

The second assumption is that the error variance is the same for all observations. This assumption may be violated because of an incorrect model specification (e.g., omitted variables, wrong functional form). Thus, it is critical to determine whether there are problems with the first assumption before we consider the second assumption. Only after we have done everything possible to avoid omitting relevant variables, or specifying inappropriate functional forms, is it useful to determine the validity of the second assumption.

A violation of the second assumption is also less critical to the interpretation of results. If the model is correctly specified but the error variance is not constant, the estimates are still unbiased.[3] The assumption of a constant variance is used only in the standard error formulas. Thus, if we assume that the variance is constant when it is not, the calculation of these standard errors is wrong.

We examine the issue more closely by assuming heteroscedasticity to exist in a specific form. To simplify the exposition, we assume the basic linear model with one predictor variable:

$$Y_i = \alpha + \beta X_i + u_i$$

Suppose that

$$E(u_i) = 0$$

and

$$\text{Var}(u_i) = \sigma^2 X_i^2$$

Note that we assume that the first assumption holds. However, instead of being constant, the error variance increases by the squared value of the predictor variable. For this specific form of heteroscedasticity we can make a correction resulting in the estimation of a different equation, one whose error term meets the second assumption and allows estimation of the parameters of interest.

If we multiply both sides of the equation by $1/X_i$, we obtain

$$\frac{Y_i}{X_i} = \frac{\alpha}{X_i} + \beta \frac{X_i}{X_i} + \frac{u_i}{X_i} \quad \text{or}$$

$$\frac{Y_i}{X_i} = \beta + \alpha \frac{1}{X_i} + \frac{u_i}{X_i}$$

In this form the error term, u_i/X_i, still has expected value equal to zero, since $E(u_i) = 0$ for a given value of X_i. The variance of this error term is

$$\text{Var}\left(\frac{u_1}{X_i}\right) = \frac{1}{X_i^2} \text{Var}(u_i) = \frac{1}{X_i^2} \sigma^2 X_i^2 = \sigma^2$$

Thus, the error term for the modified equation satisfies both the first and the second assumptions. For this specific instance of heteroscedasticity the new criterion variable is Y_i/X_i, and the predictor variable is $1/X_i$. Also, note that the modified equation's intercept is the slope estimate in the original equation. And, the slope coefficient for $1/X_i$ is the original equation's intercept.

Intuitively, the transformation involves weighting each observation by $1/X_i$ so that as the value for X_i increases, the weight declines. Thus, larger values of X_i are weighted less heavily than are smaller values. This makes sense because the amount of uncertainty as measured by the variance of the error term increases as X_i increases. Figure 7.1 shows how the observations tend to be further away from the plotted average relation between

Figure 7.1
$\text{Var}(u_i)$ increases as X_i increases

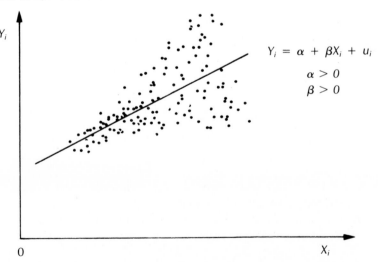

Y and X as X_i increases. The correction involves weighting the observations inversely by the amount of uncertainty, in order to estimate the parameters most efficiently. In general, however, before we make such corrections for heteroscedasticity, we should convince ourselves that the reason for the heteroscedasticity is not due to model misspecification. If we know we have done everything possible to get the correct model specification, we should be able to justify the existence of heteroscedasticity. Thus, it should be possible to anticipate a violation of the second assumption, based on our knowledge of the substantive aspects of the problem. Plots and tests can be used to verify our anticipation and, in general, to detect heteroscedasticity.

Plots

The residuals for an equation with an intercept are on the average equal to zero (see Appendix 2.3). If the pattern shown in Figure 7.1 for a simple linear model exists, then the (absolute) value of the residual tends to increase as the value for X_i increases. This is shown in Figure 7.2. In the simple model, the plot of residuals versus the observed values for the predictor variable can help us detect heteroscedasticity. In the multiple-regression model, the residuals can be plotted against the fitted values and also separately against each of the predictor variables. The former may be sufficient to detect heteroscedasticity; however, the latter plots are necessary to verify or determine which predictor variable(s), if any, are responsible for the heteroscedasticity. In terms of the

Figure 7.2
(Absolute) magnitude of residual increases as X increases

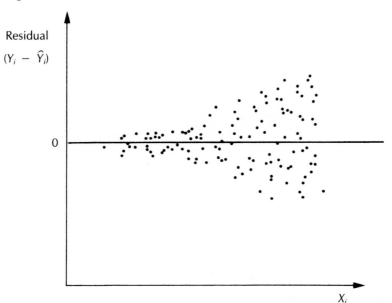

Figure 7.3
No evidence of heteroscedasticity

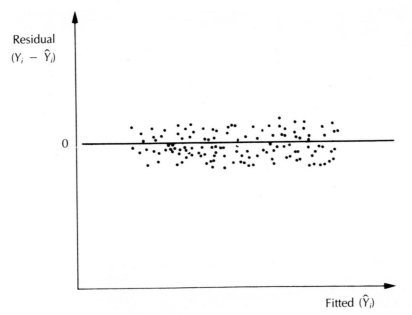

plot of residuals versus fitted values, we expect a pattern such as shown in Figure 7.3 in the *absence* of heteroscedasticity. The same pattern indicates this absence when the residuals are plotted against a given predictor variable.

Tests

Tests can be used to determine whether the deviation from constant variance is statistically significant. A test developed by Goldfeld and Quandt proceeds as follows.[4] Order the observations by the magnitude of the values for X_i, and, for the purpose of testing for heteroscedasticity only, omit some observations in the middle. In general, one fifth of the observations may be omitted. Let the median value for X_i equal Med. We omit those observations that are closest to Med, half below and half above Med. Then do two separate regression analyses:

1. Regress Y on X using the smallest X_i values (sample size equals $2/5\ n$).
2. Regress Y on X using the largest X_i values (sample size equals $2/5\ n$).

Using the residuals from each regression, compute the sample variances

$$s_1^2 = \frac{\Sigma(Y_{1i} - \hat{Y}_{1i})^2}{\left(\dfrac{2n}{5} - 2\right)}$$

$$s_2^2 = \frac{\Sigma(Y_{2i} - \hat{Y}_{2i})^2}{\left(\dfrac{2n}{5} - 2\right)}$$

where Y_{1i} refers to the observed values in the first subset (smallest X_i values), and Y_{2i} refers to the observed values in the second subset (largest X_i values).

Under the null hypothesis that $\sigma_1^2 = \sigma_2^2$ (homoscedasticity), it can be shown that

$$\frac{s_2^2}{s_1^2}$$

follows an F-distribution with $\left(\dfrac{2n}{5} - 2\right)$ and $\left(\dfrac{2n}{5} - 2\right)$ degrees of freedom.[5] If the computed F-value exceeds the appropriate tabulated value, significant evidence of heteroscedasticity exists and a correction should be made.

The test procedure discussed above can be altered slightly as follows. If we assume that the relation between Y and X is the same for all values of X, then there is no reason to estimate two separate regressions. Instead, we may estimate one regression using all the data in the sample and separate the residuals from this regression into two groups based on the value of X_i, as discussed above.

If heteroscedasticity appears to exist, and if we do not know which predictor variable can be used to make a correction, we can use plots of the residuals against each predictor variable separately. Alternatively, a regression analysis can be carried out with the *absolute* values of the residuals, $|\hat{u}_i| = |Y_i - \hat{Y}_i|$, as the criterion variable, and with $X_{1i}, X_{2i}, \ldots, X_{mi}$ (or transformations, such as X_{1i}^2 or $X_{1i}^{1/2}$) as the predictor variables. The objective of this regression is to choose the predictor variable(s) most useful for the heteroscedasticity correction.

Problems

7.1 Suppose the following model is of interest:

$$Y_i = \alpha + \beta X_i + u_i$$

There are two assumptions about the error term:

Assumption 1: $E(u_i) = 0$

Assumption 2: $\text{Var}(u_i) = \sigma^2/X_i^2$

If these assumptions are valid, find the correction for heteroscedasticity and show that the resulting error term has constant variance.

7.2 Suppose the following model is of interest:

$$Y_i = \alpha + \beta X_i + u_i$$

There are two assumptions about the error term:

$$\text{Assumption (1)} \quad E(u_i) = 0$$

$$\text{Assumption (2)} \quad \text{Var}(u_i) = \sigma^2 X_i$$

If these assumptions are valid, find the correction for heteroscedasticity and show that the resulting error term has constant variance.

Reasons for $\text{Cov}(u_i u_j) \neq 0$, for $i \neq j$

The third assumption which we made earlier is that the errors are independent for any pair of observations. We can visualize the independence by imagining the error term to be something like a coin. Having tossed the coin for one observation and observed a head as the outcome, we have no useful information to help us predict the outcome of the next coin toss. The outcomes of coin tosses are independent, and that is what we are assuming for the error term. Another way of looking at this is to say that the criterion variable contains systematic variation (the outcomes of the criterion variable may not be independent), and our model is supposed to capture this systematic variation. If the model is adequate, the unexplained variation should exhibit a pattern similar to the outcomes of coin tosses, in the sense that the outcomes are independent of each other.

The assumption of independent errors may be violated as a result of shortcomings in the model specification. For example, if a relevant predictor variable is omitted or an incorrect functional form is used, the errors may not be independent. Thus, just as with the second assumption, it is critical that we have satisfied ourselves to the fullest extent with regard to model specification. Indeed, if we encounter a violation of independence, we have to identify plausible reasons. Under certain conditions a lack of independence for the errors is anticipated (for example, see Chapter 9). In general, it is impossible to state what action to take, when the assumption is violated, without knowing the reason for the violation.

A violation of the third assumption, if it is not the result of model misspecification, is not critical. As long as the first assumption holds, the parameter estimates are unbiased. The third assumption, just like the second, is used only to obtain a simple formula for the standard errors of the parameter estimates. If the third assumption is violated, the calculation of these standard errors is wrong. The variables can be transformed such that the assumption of independence holds for the model with the transformed variables. In practice, however, we cannot claim that we only have to transform the variables if we determine that the third assumption is violated. Implicitly, this transformation to obtain independence assumes that the model is correctly specified. Unfortunately, a lack of independence often results from model misspecification.

We first present the argument in its traditional form. The typical treatment is to consider a sample of time series data.[6] We use a simple linear model for illustration:

$$Y_t = \alpha + \beta X_t + v_t$$

where v_t represents the error term in this model, and

t represents the time period.

Suppose that $E(v_t) = 0$ for all t, and $v_t = \rho v_{t-1} + e_t$. In other words, the error terms are not independent but follow a first-order autoregressive process with parameter ρ (rho). Assume also that $E(e_t) = 0$, and $E(e_t, e_s) = 0$ for $t \neq s$. Then it is appropriate to transform the variables such that we obtain a different equation whose error term meets the third assumption. Given our assumptions about v_t, it is easy to see that the different equation should have e_t as the error term. This is accomplished when we use

$$Y_t - \rho Y_{t-1} = \alpha(1 - \rho) + \beta(X_t - \rho X_{t-1}) + e_t$$

Note that this formulation has considerable similarity with the use of first differences. But instead of using $(Y_t - Y_{t-1})$ and $(X_t - X_{t-1})$ as the variables, lagged values are multiplied by the autocorrelation parameter ρ. Also, the slope parameter for the transformed predictor variable is identical to the one in the original equation. However, the intercept is a combination of the original intercept and the autoregressive parameter that characterizes the lack of independence in v_t, the original error term.

The discussion above suggests that it is possible for the first assumption to hold while the third assumption is violated. Unfortunately, many people have taken this to mean that the parameter estimates are still unbiased if the third assumption is violated. Of course, this is not true if the reason for the violation is a misspecification of the model. More often than not, a violation of the third assumption is due to omitted variables or functional form inadequacies. The use of a transformation to correct for a lack of independent error terms is justified only if the model specification is critically evaluated and a justification for the autocorrelation in the error term is provided.

Plots

With time-series data, a lack of independence in the error term exists if there is a systematic pattern in the behavior of the residuals over time. We assume that time is not a predictor variable in the model (as in the example of AT&T dividends as a function of trend in Chapter 3). A scatterplot of the residuals against time can thus be used to see if they are autocorrelated. Figures 7.4 and 7.5 show two distinct cases of positive (7.4) and negative (7.5) autocorrelation.

In Figure 7.4, we see that all of the first twenty or so observations have negative residuals, while the last twenty or so are all positive. Positive autocorrelation means that the residual in period t tends to have the same sign as the residual in period $(t - 1)$. It can

Figure 7.4
Positive autocorrelation

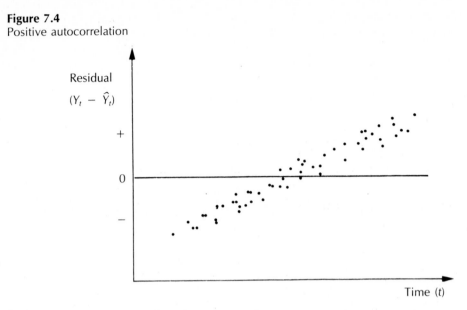

be measured by correlating the residuals $(\hat{u}_t)$ with the same residuals lagged one period $(\hat{u}_{t-1})$.

Similarly, we see in Figure 7.5 that the observations tend to have positive values followed by negative ones, and vice versa. Thus, negative correlation means that the residual in period t tends to have the opposite sign as the residual in period $(t - 1)$, as we would observe by correlating the residuals $(\hat{u}_t)$ with the lagged residuals $(\hat{u}_{t-1})$.

Figure 7.5
Negative autocorrelation

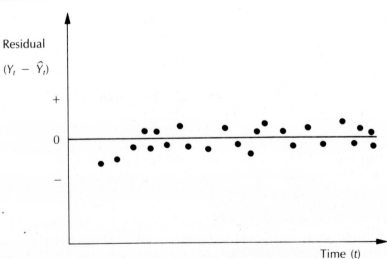

Tests

From the discussion above, it is easy to conceive of one possible index for autocorrelation—namely, the correlation coefficient between the residuals and the corresponding lagged residuals:

$$r = \frac{\displaystyle\sum_{t=2}^{n} \hat{u}_t \hat{u}_{t-1}}{\displaystyle\sum_{t=2}^{n} \hat{u}_{t-1}}$$

However, the most widely used index and test for first-order autocorrelation is the Durbin-Watson statistic[7]:

$$d = \frac{\displaystyle\sum_{t=2}^{n} (\hat{u}_t - \hat{u}_{t-1})^2}{\displaystyle\sum_{t=1}^{n} \hat{u}_t^2}$$

It can be shown that under the null hypothesis of no autocorrelation the expected value of d is approximately

$$E(d) \cong 2 + \frac{2(m-1)}{n-m}$$

For convenience, we assume the expected value of d to be approximately 2 (this is reasonable especially if m is small relative to n). If the calculated value for d is below 2, the residuals tend to be positively autocorrelated. If d exceeds 2, the first-order autocorrelation tends to be negative. Critical values for rejection of the null hypothesis are shown in Table V. The entries in this table depend on the number of predictor variables (m), the sample size (n), and the type I error probability. However, even then there are two entries: d_L (lower bound) and d_U (upper bound). The rules for determining whether the computed d-value is statistically significant are

- If $d < d_L$, reject the null hypothesis of no first-order autocorrelation and conclude there is evidence of positive autocorrelation.
- If $d_U < d < 2$, do not reject the null hypothesis of no first-order autocorrelation.
- If $d_L < d < d_U$, the Durbin-Watson test is inconclusive. We cannot reject the null hypothesis, even though there may be positive autocorrelation.
- If $d < 2$, compute $d' = (4 - d)$ and use the same rules above, but substitute d' for d, and negative for positive autocorrelation.

Based on the value for d, the first-order autocorrelation coefficient can be estimated as follows:

$$\hat{\rho} = \frac{2 - d}{2}$$

An alternative procedure for checking first-order autocorrelation involves the *runs test*. This test disregards the residuals' magnitudes and uses only their signs (positive or negative). A run is a series of consecutive residuals with identical signs. For example, suppose we observe the signs of a sample of ten residuals as follows:

$$
\begin{array}{ccccc}
1 & 2 & 3 & 4 & 5 \\
\overline{|+\;+|} & \overline{|-|} & \overline{|+|} & \overline{|-\;-\;-|} & \overline{|+\;+\;+|}
\end{array}
$$

In this sample there are six positive ($+$) and four negative ($-$) residuals. Note that if the model contains an intercept, the average value of the residuals equals zero. Hence, the residuals cannot be all positive or all negative. However, the number of positive residuals does not have to be equal to the number of negative residuals. The first run consists of two positive residuals, the second has one negative residual, the third run has one positive residual, and so forth. The question the runs test tries to answer is whether the observed number of runs deviates significantly from the expected number of runs. The null hypothesis is that the error terms are independent, or specifically that consecutive error terms are independent. Thus, like the Durbin-Watson test, the runs test can be used only to investigate first-order autocorrelation. First-order means that there is a lack of independence between the errors one period apart.

To determine statistical significance we need to know the frequency distribution for the possible number of runs under the null hypothesis. This frequency distribution depends on the sample size (number of residuals) and the proportion of positive and negative residuals observed in the sample. For a sample size of ten the smallest possible number of runs is two (for example, one run of positive residuals followed by one run of negative residuals). The largest possible is ten (each time a positive residual occurs it is followed immediately by a negative residual, and vice versa). However, if there are nine positive and one negative residuals, the largest possible number of runs equals three.

Tables have been constructed to determine whether an observed number of runs differs significantly from what would be expected under the null hypothesis. These tables usually indicate the extreme outcomes (extremely small or extremely large number of runs) required for rejection of the null hypothesis at a given type I error probability. For example, if the sample size equals ten and there are five positive and five negative residuals, the null hypothesis is rejected at the five-percent level if the number of runs equals two or ten. Thus, if we have the outcomes of ten coin tosses, we can reject the null hypothesis of independence only if we get two runs (e.g., five heads followed by five tails) or ten runs (e.g., a head is always followed by a tail, and vice versa).

Sample sizes in regression analysis usually exceed the sizes assumed in most published tables for a runs test; however, for large samples the normal distribution can be used as an approximation. The test statistic is

$$Z = \frac{r - \mu_r}{\sigma_r}$$

where r is the observed number of runs,

   μ_r is the expected number of runs under the null hypothesis, and

   σ_r is the standard deviation in the number of runs.

The formulas for μ_r and σ_r are[8]

$$\mu_r = \frac{2n_1 n_2}{n_1 + n_2} + 1$$

$$\sigma = \sqrt{\frac{2n_1 n_2(2n_1 n_2 - n_1 - n_2)}{(n_1 + n_2)^2(n_1 + n_2 - 1)}}$$

where n_1 and n_2 are the number of positive and negative residuals in the sample.

Note that if $r < \mu_r$, so that $Z < 0$, the residuals tend to be positively autocorrelated, while if $r > \mu_r$, so that $Z > 0$, they tend to be negatively autocorrelated. Statistical significance is indicated when the computed Z-value exceeds the cutoff value for the desired type I error probability.

Autocorrelation Due to Model Misspecification

We use the example of marginal cost and cumulative production (see Chapter 6) to illustrate how model misspecification may result in a lack of independence in the errors. A linear relationship, estimated from the data in Table 6.3, results in the equation

$$\hat{Y}_t = 70.5 - 4.5X_t$$

where $\hat{Y}_t$ is the predicted marginal unit cost

   X_t is the cumulative production

We have argued in Chapter 6 that this linear equation is not adequate. Indeed, the Durbin-Watson statistic is 0.71, which is statistically significant at the five-percent level. We conclude that there is evidence of first-order positive autocorrelation in the error term. We cannot claim, though, that the parameter estimates are unbiased. In this case we know that the functional form used is incorrect. Of course, the autocorrelation will disappear once the appropriate functional form is used. However, the transformation involving the lagged values of criterion and predictor variables (see below) is inappropriate.

Conventional Autocorrelation Treatment

Assuming that autocorrelation is not due to model misspecification, we can proceed as follows. Let

$$Y_t = \alpha + \beta X_t + v_t \qquad t = 1, 2, \cdots, n$$

$$v_t = \rho v_{t-1} + e_t$$

where ρ is the autocorrelation parameter.

Also,
$$|\rho| < 1$$

$$E(e_t) = 0 \text{ for all } t$$

$$E(e_t e_s) = \sigma_e^2 \text{ if } t = s$$
$$= 0 \text{ if } t \neq s$$

Under these assumptions, it can be shown that the least-squares estimator is unbiased but inefficient. If, in fact, it is used to estimate β, the standard formula for the variance of the OLS estimator (see Appendix 7.1) is incorrect, that is,

$$\text{var } \hat{\beta} \neq \frac{\sigma_v^2}{\Sigma x_t^2}$$

where $x_t = (X_t - \bar{X})$.

The correct formula is

$$\text{var } \hat{\beta} = \frac{\sigma_v^2}{\Sigma x_t^2} \left[1 + 2\rho \frac{\Sigma x_t x_{t+1}}{\Sigma x_t^2} + \cdots + 2\rho^{n-1} \frac{x_1 x_n}{\Sigma x_t^2} \right]$$

A comparison of these formulas shows that the bias inherent in the incorrect formula is small if X does not contain a trend. However, if v_t is positively autocorrelated ($\rho > 0$) and X follows a trend, then the incorrect formula underestimates the variance. Furthermore, if we use

$$\hat{\sigma}_v^2 = \frac{\Sigma \hat{v}_t^2}{n - 2}$$

it can be shown that $E(\hat{\sigma}_v^2) < \sigma_v^2$.

We conclude, from the conventional treatment of autocorrelation, that if we can assume that our model is well-specified (i.e., $E(v_t) = 0$ for all t), the least-squares estimator is still unbiased. However, the formula for the variance (and hence the standard error) of the slope coefficient is biased downward if v_t is positively autocorrelated and X_t follows a trend. We discuss remedies for autocorrelation (if it is not due to model misspecification) in Appendix 7.1.

Reasons for u_i Not Being Normally Distributed

By now it should be no surprise that the fourth assumption may also be violated due to model misspecification. Again, we suggest that this assumption should not be examined until the model specification is satisfactory and the second and third assumptions have been checked as well. The fourth assumption is needed to justify the use of a particular distribution to test hypotheses and to construct confidence intervals. We can examine the validity of this assumption indirectly through the residuals. If each error term separately satisfies all four assumptions, then the estimated error values (residuals) as a group will be normally distributed. Thus, if the residuals appear to be normally distributed, we cannot reject the hypothesis that each unobservable error term follows a normal distribution.

Plots

To examine whether the residuals are approximately normally distributed, we can classify the residuals into groups and construct a histogram. For example, if we decide to have eight classes, we may group the standardized residuals as follows (the residuals are standardized by dividing the observed values by the standard deviation of the residuals):

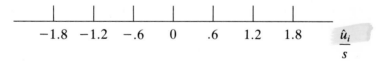

The histogram will show the relative frequency with which the standardized residuals fall into each of the eight classes. If the errors are normally distributed, we expect the following result (the expected relative frequencies can be obtained directly from a standard normal distribution table, using the class boundaries as Z-values):

Lower class boundary	Upper class boundary	Expected relative frequency (%)
$-\infty$	-1.8	3.5
-1.8	-1.2	8.0
-1.2	-0.6	16.0
-0.6	0	22.5
0.6	1.2	16.0
1.2	1.8	8.0
1.8	∞	3.5
		100.0

If an uneven number of classes is desired, we could construct seven groups:

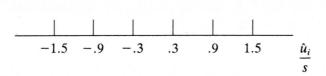

For this arrangement we expect, under the null hypothesis of normally distributed errors, the following relative frequency distribution for the standardized residuals:

Lower class boundary	Upper class boundary	Expected relative frequency (%)
$-\infty$	-1.5	6.5
-1.5	-0.9	12.0
-0.9	-0.3	20.0
-0.3	0.3	23.0
0.3	0.9	20.0
0.9	1.5	12.0
1.5	∞	6.5
		100.0

Based on the histogram chosen, we look for systematic deviations from normality. Such deviations may occur in the form of asymmetry (skewed distributions). We show in Figure 7.6 an example of a distribution that is positively skewed (skewed to the right). With a positive skew, the measures of central tendency are located from left to right along the horizontal axis as follows: mode, median, mean. For example, $\mu > $ Med indicates a positively skewed distribution.

Figure 7.6
A postively skewed distribution

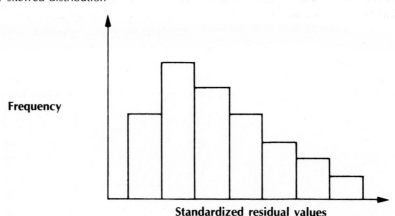

Figure 7.7
A negatively skewed distribution

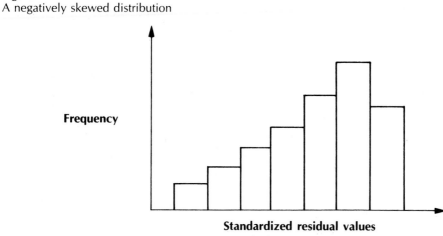

A negatively skewed (skewed to the left) distribution is shown in Figure 7.7. With a negative skew, the measures of central tendency are located from left to right along the horizontal axis as follows: mean, median, mode. In this case, $\mu <$ Med indicates a negatively skewed distribution.

Tests

A relatively simple test for normality involves the chi-square distribution. This test compares the observed frequencies for a given set of categories with the frequencies expected under the null hypothesis of normally distributed error terms. The test is most powerful if the boundaries of the categories are determined in such a manner that each category has the same expected frequency. For instance, if we decide to use five classes, we want to obtain boundaries (Z-values) such that the expected relative frequencies are 20 percent for each class. In terms of standardized residuals, the boundaries are the following:

Lower class boundary	Upper class boundary	Expected relative frequency (%)
$-\infty$	-0.84	20
-0.84	-0.25	20
-0.25	0.25	20
0.25	0.84	20
0.84	∞	$\underline{20}$
		100

For large sample sizes, we may prefer to work with a larger number of classes to make maximum use of the data. For example, if we use eight classes we obtain the boundaries shown below:

Lower class boundary	Upper class boundary	Expected relative frequency (%)
$-\infty$	-1.15	12.5
-1.15	-0.67	12.5
-0.67	-0.32	12.5
-0.32	0	12.5
0	0.32	12.5
0.32	0.67	12.5
0.67	1.15	12.5
1.15	∞	12.5
		100.0

The test involves the following statistic

$$\chi^2 = \sum_{j=1}^{J} \frac{(O_j - E_j)^2}{E_j}$$

where O_j is the observed frequency in category j

E_j is the expected frequency in category j (expected relative frequency multiplied by the total number of residuals in the sample)

J is the number of categories

The computed chi-square value is then compared with the tabulated value for a specified type I error probability and $(J - 1 - 2)$ degrees of freedom.[9]

Other Diagnostics

Recall that least-squares analysis minimizes the sum of the squared deviations between observed and fitted values for the criterion variable. The estimated equation is therefore sensitive to extreme observations, just as the mean value of a set of observations is sensitive to extreme observations. Many regression software packages provide warnings when such observations are present. They are often referred to as *outliers*.

An outlier is defined in terms of the residual values remaining after estimating the regression equation. Specifically, an outlier is an observation for which the actual value of Y differs greatly from the fitted value as compared with the other residuals. An outlier could be defined somewhat arbitrarily as any observation for which the residual value is more than two standard deviations away from zero. Due to the extraordinarily large effect such an observation can have on the regression result, it is useful to identify outliers. However, the existence of one or more outliers does not imply that an assumption of the model has been violated.

If outliers exist, the following procedure may be adopted. First, check the data. Perhaps a recording or typing error has been made, and as a result one or more observations have unusually large residuals. Obviously, if some part of the data is incorrect, it should be corrected before continuing with the analysis. Second, if there are no data problems, check whether there are unusual events which are not taken into account in any of the predictor variables. There are several options for dealing with such a situation. For example, such unusual events could be incorporated into the set of predictor variables. Or, if they cannot be incorporated in a reasonable manner, it may be appropriate to omit the affected observations from the data. Finally, it may be useful to present results both with and without outliers. We can then make a choice about the appropriate result to use, based on our knowledge of the circumstances that cause outliers to exist.

An alternative definition of outliers exists, based on a comparison of the values of the predictor variables. For example, for a given predictor, it is useful to know whether there are observations that have values, say, more than two standard deviations away from the sample mean of that predictor variable. Again, the existence of such outliers does not indicate that an assumption is violated. Instead, the information is useful because it indicates that there may be observations with substantial influence on the estimated relationship. If the reason is a data error, the problem is easily corrected. If it is not a data error, it is useful to consider the reason for the existence of outliers in terms of the predictor, and the influence these outliers have on the observed result. For example, we can temporarily delete such outliers from the sample and compare the result after deletion with the one before deletion. The difference is an indication of the influence these outliers have.

SUMMARY

In this chapter, we have discussed the importance of investigating the validity of assumptions made about the error term. By far the most critical assumption is the first one. Unfortunately, our ability to test this assumption, that the expected value of the error term is zero, is quite limited. If there are many shortcomings in the model (e.g., omitted variables, functional form problems), residual plots and tests may indicate a deficiency but are unlikely to pinpoint the required correction. Other factors that play a role in the value of diagnostic procedures include the correlation between included and omitted predictors, and the range of sample variation for the included predictors.

Assuming that we have maximized the ability to develop appropriate model specifications before data analysis, we can use both residual plots and model tests to see if the data are consistent with the error assumptions. Only if there is no evidence that the first assumption about the error term is violated or if the model is corrected is it of interest to consider the other assumptions. The reason for considering assumptions one at a time and in a particular order is not just that the first assumption is the most critical one. It is also because a violation of the first assumption is extremely likely to show up as a violation of one or more other assumptions as well.

For each assumption we have discussed some of the available plots and tests. If the model used in an empirical study is well-specified, there should be no evidence that one or more assumptions are violated. Nevertheless, if there is no evidence of a violation, we cannot be sure that the model is indeed well-specified. We may not have enough data, for example, or we may have insufficient variation in a predictor variable to find enough evidence against an assumption.

Interpretation of a violation requires substantive knowledge about the problem under study. For instance, if the residuals show evidence of heteroscedasticity, there must be an acceptable, logical reason for it. With cross-sectional data, such reasons may include different degrees of (sampling) error for the observations for the criterion variable; we may also have persuasive arguments for why the magnitude of the error should be related to the value of the criterion variable (or a predictor variable). Similarly, for autocorrelated errors, there must be a valid argument. Often, a finding of autocorrelation results in a modification of the estimation method (such as the Cochrane-Orcutt procedure in Appendix 7.1[10]). However, this technical or statistical adjustment is not useful if the reason for the autocorrelation is a mistake in the model specification.

In the final section of this chapter, we mentioned the need to consider the existence of outliers in the data. Although they are not related to the assumptions we made about the error term, it is important to identify outliers. For example, a large (negative or positive) residual may result from a mistake in data input. If not, it may correspond to an observation that occurs under unusual circumstances. In that case, the identification of outliers may prompt a modification in the model. It is also common to consider the existence of outliers in the predictor variables. For example, if a sample includes many observations for relatively small values of a predictor, and one for which the same predictor has a very large value, this one observation would determine the value of the predictor's slope coefficient. No regression results should be used without comprehensive consideration of the issues discussed in this chapter.

ADDITIONAL PROBLEM

7.3 The Brisk Company manufactures and sells young-women's fashion clothing to retailers. Data for 18 consecutive quarters have been collected for the purpose of predicting the Brisk Company's revenues. The basic variables of interest are

REV   = revenues for Brisk in thousands of current dollars for a given quarter,

WRTW   = total retail sales for all women's ready-to-wear clothing in the U.S. in millions of current dollars for a given quarter, and

INDEX = seasonally adjusted price index of women's and girls' apparel (United States city average).

After a discussion with management, the researcher in charge of model development decided to start with the following equation:

$$REV_t = \beta_0 + \beta_1 WRTW_t + \beta_2 INDEX'_t + \beta_3 Q2_t + \beta_4 Q3_t + \beta_5 Q4_t + u_t$$

where   Q2 = 1 if the observation is for the second quarter of the calender year

= 0 otherwise

Q3 = 1 if the observation is for the third quarter of the calender year

= 0 otherwise

Q4 = 1 if the observation is for the fourth quarter of the calender year

= 0 otherwise

Brisk Company data

ROW	REV	WRTW	INDEX	Q2	Q3	Q4
* 1 *	1977.2	2011.0	120.4	0	1	0
* 2 *	1855.5	2184.0	121.3	0	0	1
* 3 *	1429.3	1938.0	122.4	0	0	0
* 4 *	3474.7	1888.0	122.9	1	0	0
* 5 *	5508.0	2195.0	122.5	0	1	0
* 6 *	6698.7	2369.0	124.4	0	0	1
* 7 *	6917.8	2223.0	124.4	0	0	0
* 8 *	9050.9	2119.0	127.0	1	0	0
* 9 *	9542.8	2355.0	128.1	0	1	0
* 10 *	9033.8	2446.0	129.6	0	0	1
* 11 *	7326.9	2271.0	130.8	0	0	0
* 12 *	11071.9	2324.0	133.4	1	0	0
* 13 *	10882.0	2535.0	137.3	0	1	0
* 14 *	10000.5	2661.0	138.2	0	0	1
* 15 *	10774.9	2516.0	136.8	0	0	0
* 16 *	11378.8	2580.0	136.5	1	0	0
* 17 *	14836.5	2880.0	139.5	0	1	0
* 18 *	17135.1	3061.0	139.5	0	0	1

```
VARIABLE        MEAN        STD. DEV.

WRTW        2.36422E+03   3.09638E+02
INDEX       1.29722E+02   6.89455E+00
Q2          2.22222E-01   4.27793E-01
Q3          2.77778E-01   4.60889E-01
Q4          2.77778E-01   4.60889E-01
REV         8.27197E+03   4.34862E+03

BASED ON 18  ACTIVE ROWS.
```

Correlation Matrix

```
             REV      WRTW      INDEX      Q2        Q3        Q4

REV       1.000
WRTW      0.902     1.000
INDEX     0.903     0.875     1.000
Q2        0.060    -0.242     0.018     1.000
Q3        0.041     0.064    -0.015    -0.331     1.000
Q4        0.099     0.371     0.081    -0.331    -0.385     1.000
```

Criterion variable: REV

```
VARIABLE         B         STD. ERROR(B)      T

WRTW        1.4192E+01    3.4684E+00       4.092
INDEX       2.0095E+01    1.4119E+02       0.142
Q2          2.2360E+03    9.9362E+02       2.250
Q3         -.3274E+03     1.0121E+03      -0.324
Q4         -.2068E+04     1.2225E+03      -1.691
CONSTANT   -.2772E+05     1.1213E+04      -2.472
```

```
                MULTIPLE R    R-SQUARE

UNADJUSTED        0.9643        0.9299
ADJUSTED          0.9491        0.9007

STD. DEV. OF RESIDUALS = 1369.99

DURBIN-WATSON STAT. = 0.880
```

a. Interpret the correlation coefficients between REV and Q2, REV and Q3, and REV and Q4.

b. The coefficients for the indicator variables Q3 and Q4 in the regression equation are negative. Why are these signs not the same as the corresponding signs in the correlation matrix?

c. The computer output also includes several graphs (see figure 7.8 and 7.9) and a Durbin-Watson statistic. Considering all these diagnostics and the regression results, provide an outline of the steps you would follow to obtain a final model.

Figure 7.8
Scatterplot of 18 standardized values of 'REV' vs. 'WRTW'

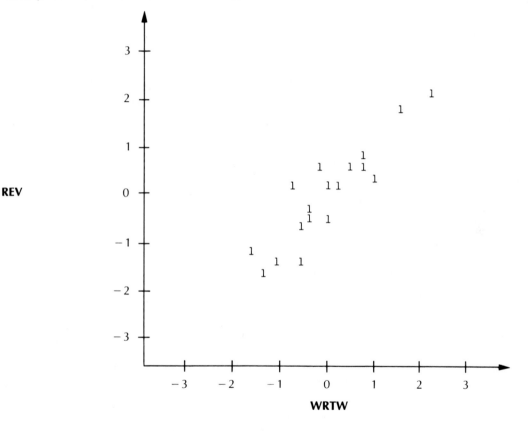

	MEAN	STD. DEV.
VERT. VAR.	8271.97	4348.62
HORIZ. VAR.	2364.22	309.638
SAMPLE SIZE = 18		

Figure 7.9
Scatterplot of 18 standardized values of 'RESIDU' vs. 'FITTED'

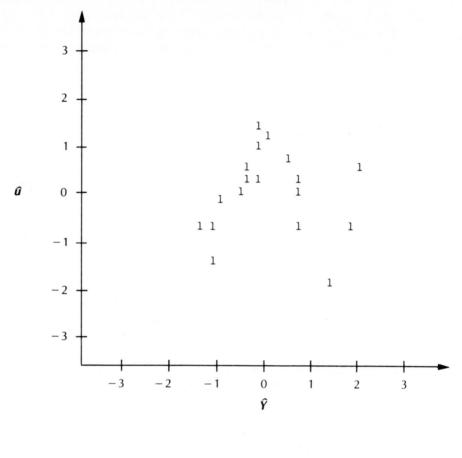

	MEAN	STD. DEV.
VERT. VAR.	0	1369.99
HORIZ. VAR.	8271.97	4193.52
SAMPLE SIZE = 18		

APPENDIX 7.1
REMEDIES FOR AUTOCORRELATION

Suppose that we believe our model is well-specified but that the error term follows a first-order autoregressive process. Then it is desirable to make a correction for the autocorrelation. The Cochrane-Orcutt procedure is commonly used and, for known values of ρ, amounts to the following transformation. Let Y^1 be a $(n-1)$ column vector and X^1 be a $(n - 1) \times 2$ matrix. Define

$$
Y^1 = \begin{bmatrix} Y_2 - \rho Y_1 \\ Y_3 - \rho Y_2 \\ \vdots \\ Y_n - \rho Y_{n-1} \end{bmatrix} \qquad X^1 = \begin{bmatrix} 1 - \rho & X_2 - \rho X_1 \\ 1 - \rho & X_3 - \rho X_2 \\ \vdots & \vdots \\ 1 - \rho & X_n - \rho X_{n-1} \end{bmatrix}
$$

We call the regression of Y^1 as a function of X^1 the *Cochrane-Orcutt (COP) estimator*.

An alternative estimator is to use Y^0, a column vector of length n, and X^0, a $(n \times 2)$ matrix, defined below:

$$
Y^0 = \begin{bmatrix} \sqrt{1 - \rho^2}\, Y_1 \\ Y_2 - \rho Y_1 \\ \vdots \\ Y_n - \rho Y_{n-1} \end{bmatrix} \qquad X^0 = \begin{bmatrix} \sqrt{1 - \rho^2} & \sqrt{1 - \rho^2}\, X_1 \\ 1 - \rho & X_2 - \rho X_1 \\ \vdots & \vdots \\ 1 - \rho & X_n - \rho X_{n-1} \end{bmatrix}
$$

We call the regression of Y^0 as a function of X^0 the *Generalized Least-Squares (GLS) estimator*. Note that the difference between GLS and COP consists of the extra first row (unit) in the Y^0 vector and the X^0 matrix. And, let the term *OLS* stand for the *Ordinary Least-Squares* (OLS) estimator, described in Chapter 3.

To compare the three estimators, let $\text{var}(\hat{\beta}_{GLS})$, $\text{var}(\hat{\beta}_{COP})$, and $\text{var}(\hat{\beta}_{OLS})$ be the true variances of the slope estimators. For the comparisons, we continue to assume that ρ is known. It can be shown that the OLS estimator is inefficient, compared to GLS, under the conditions specified above. Thus,

$$
\frac{\text{var}(\hat{\beta}_{OLS})}{\text{var}(\hat{\beta}_{GLS})} > 1 \text{ for } |\rho| \neq 0
$$

However, for positive autocorrelation $(\rho > 0)$, and X_t following a linear trend, the following has been shown[11]:

$$
\frac{\text{var}(\hat{\beta}_{OLS})}{\text{var}(\hat{\beta}_{COP})} \to 0 \text{ as } \rho \to +1
$$

Thus, under the conditions we identified earlier as causing serious problems for the OLS estimator (X_t following a trend, $\rho > 0$), the Cochrane-Orcutt procedure is likely to be inefficient relative to OLS.

Intuitively, the potential problem with COP can be seen as follows. If X_t follows a linear trend and ρ is close to 1, then the second column in X^1 is almost without variation. Thus, the collinearity between the first and second columns in X^1 becomes very high. This is not the case for GLS, since X^0 contains one additional row of observations. When ɔrmation is chosen, these aspects should be considered. We show in ⱴ the Cochrane-Orcutt procedure can be used with an estimate of ρ.

APPENDIX 7.2
AUTOCORRELATION ADJUSTMENT

Procedure to investigate autocorrelation, and a mechanical correction Assumed model:

$$Y_t = \alpha + \beta_1 X_{1t} + \beta_2 X_{2t} + \cdots + v_t$$

Suppose the Durbin-Watson statistic is significant. Let

$$\hat{\rho} = \frac{2 - d}{2}$$

Corrected model:

$$Y_t - \hat{\rho} Y_{t-1} = \alpha(1 - \hat{\rho}) + \beta_1(X_{1,t} - \hat{\rho} X_{1,t-1}) + \beta_2(X_{2,t} - \hat{\rho} X_{2,t-1}) + \cdots + e_t$$

This represents the Cochrane-Orcutt procedure.
   Now, for the error term e_t

1. The Durbin-Watson statistic is *not* significant. Usually if there is an autocorrelation problem, it is limited to first-order autocorrelation; that is, if the error v_t in the assumed model shows a violation of the independence assumption, the error e_t in the corrected model does not show a violation of the independence assumption.
2. The Durbin-Watson statistic is again significant (i.e., e_t *also* shows a violation).

Procedure (if the second condition is true):

Obtain d^* from the corrected model and let

$$\hat{\rho}^* = \frac{2 - d^*}{2}$$

Twice-corrected model:

$$(Y_t - \hat{\rho} Y_{t-1}) - \hat{\rho}^*(Y_t - \hat{\rho} Y_{t-1})_{-1} = \cdots$$

$$Y_t - \hat{\rho} Y_{t-1} - \hat{\rho}^* Y_{t-1} + \hat{\rho}^* \hat{\rho} Y_{t-2} = \cdots$$

$$Y_t - (\hat{\rho} + \hat{\rho}^*) Y_{t-1} + \hat{\rho}^* \hat{\rho} Y_{t-2} = \cdots$$

Thus, if d^* is significant, the estimate of the first-order autocorrelation coefficient is $(\hat{\rho} + \hat{\rho}^*)$ and the estimate of the second-autocorrelaton coefficient is $(\hat{\rho}^* \hat{\rho})$.

ENDNOTES

1. See Churchill, Gilbert A., *Marketing Research: Methodological Foundations,* 2d ed. (Hinsdale, Ill.: The Dryden Press. 1979), p. 562.

2. We also note that the null hypothesis cannot be rejected at the one percent level of significance, where the critical F-value is 5.09. The choice of significance level introduces a notion of arbitrariness in this framework.

3. That is, we assume that the reason for the heteroscedasticity is not an omitted variable or functional form misspecification.

4. S. M. Goldfeld and R. E. Quandt. "Some Tests for Heteroscedasticity," *Journal of the American Statistical Association,* 60 (1965): 539–547.

5. The ratio is stated with s_2^2 in the numerator on the assumption that the subset based on the higher X_i values may have a larger sample variance than the other subset.

6. Time-series data are observations gathered over several time periods, such as data on railway traffic in the U.S. over thirty years. Alternatively, we may have cross-sectional data: observations for a given point in time, gathered across several cross-sections. An example of this would be observations on a variety of variables dealing with economic activity in 1986 for each of the fifty states in the U.S.

7. J. Durbin and G. S. Watson, "Testing for Serial Correlation in Least Squares Regression," *Biometrika,* 38 (1951): 159–177.

8. See, for example, Siegel, Sidney, *Nonparametric Statistics for the Behavioral Sciences* (New York: McGraw-Hill, 1956).

9. The degrees of freedom equals $(J - 1 - 2)$. The last two degrees of freedom are lost because a mean and a standard deviation are estimated for the residuals before the distributional test is performed.

10. D. Cochrane and G. H. Orcutt, "Application of Least Squares Regression to Relationships Containing Autocorrelated Terms," *Journal of the American Statistical Association,* 44 (1949): 32–61.

11. Kramer, Walter, "Note on Estimating Linear Trend When Residuals are Autocorrelated," *Econometrica,* 50 (July 1982): 1065–7.

REFERENCES

Churchill, Gilbert A. *Marketing Research: Methodological Foundations.* 2d ed. Hinsdale, Ill.: The Dryden Press, 1979.

Cochrane, D., and G. H. Orcutt, "Application of Least Squares Regression to Relationships Containing Autocorrelated Terms." *Journal of the American Statistical Association,* 44 (1949): 32–61.

Durbin, J. and G. S. Watson. "Testing for Serial Correlation in Least Squares Regression." *Biometrika,* 38 (1951): 159–177.

Goldfeld, S. M., and R. E. Quandt. "Some Tests for Heteroscedasticity." *Journal of the American Statistical Association,* 60 (1965): 539–547.

Kramer, Walter. "Note on Estimating Linear Trend when Residuals are Autocorrelated." *Econometrica,* 50 (July 1982): 1065–7.

Siegel, Sidney. *Nonparametric Statistics for the Behavioral Sciences.* New York: McGraw-Hill, 1956.

8

Relative Measures of Fit and Explanatory Power: R^2 and Beta Weights

In empirical work based on regression analysis, much attention is focused on certain summary results. For example, a model's R^2 (unadjusted or adjusted) value is often used as a basis for determining the quality and usefulness of the results. In this chapter we discuss why, and under what conditions, such an emphasis on the R^2 value may be misleading. We show how the value of R^2 can be manipulated without changing the substantive results (e.g., the slope coefficients). But we also suggest that sometimes a model with a lower R^2 value should be favored over a model with a higher R^2 value. The purpose of this chapter is to make a regression user sensitive to these issues, and to argue that the usefulness of a regression result should often be determined based on measures other than the R^2 value.

After this discussion, we present the concept of standardized regression coefficients, or *beta weights*. These standardized measures are sometimes used to infer the relative importance of each of the predictor variables in a model. They have played a role in studies where the predictor variables are defined in noncomparable units. For example, one predictor variable may be measured in dollars, while another predictor variable is measured in square feet. In such cases, if we want to compare the "importance" of the predictor variables, we cannot rely upon the slope coefficients. Essentially, beta weights are estimates of the explanatory power provided by each of the predictors. Simplified, these measures estimate the contribution to the R^2 value separately for each predictor, based on the multiple regression result. Consequently, whatever limitations are associated with the R^2 measure may also apply to beta weights. In addition, there are other difficulties encountered in the use of beta weights. We end this chapter by describing the conditions under which beta weights may be useful.

Favoring the Model with a Lower R^2

Recall the simple regression examples involving AT&T dividends, introduced in Chapter 3. The criterion variable is defined as dividends per common share, measured over the period 1973 through 1980. The first simple regression analysis uses earnings per common share in the previous year as the predictor variable. The second simple regression analysis uses a trend variable (1973 = 1, 1974 = 2, etc.) as the predictor variable. The R^2 values for the two simple regression analyses are 0.95 and 0.98, respectively. A comparison of R^2 values favors the second model.

Even though the second model provides the higher goodness-of-fit measure, we favor the first model. The main reason for this is that dividend payments are likely to depend on earnings. That is, the higher the earnings per share in one year, the more a firm can afford to pay in dividend per share in the following year. On the other hand, there is no logical reason why a firm should continue to increase dividends in the case of a positive trend, year after year.

In fact, there are several reasons why we may prefer the first model over the second. First, it provides insight as to how dividends may be determined. The trend model does not provide any information about the basis for dividend decisions. Second, if the assumed causal link between dividends and earnings exists, the first model estimates the effect of a change in earnings on dividends. Third, if earnings become depressed or decrease relative to a previous year, the predicted dividends will also be depressed or decrease.

The sample data for AT&T show a healthy growth in earnings per share over the eight-year period. The amount of dividends paid also has grown substantially. One thing the variables have in common is the amount of inflation during that period (it would be useful to express both variables in constant dollars). However, if the growth in earnings declines after the sample period, the growth in dividends should also decline. The trend model would in that case overpredict the dividends. Thus, we expect that with systematic changes in earnings per share, the first model will outperform the second model, even though it has a lower R^2. Indeed, if we make dividend forecasts for the next three years, we find that the first model outperforms the second model by a substantial margin.

Absolute versus Relative Fit

In the previous example, both models have relatively high R^2 values, and we might believe that there is not a great deal to choose between in those two models, in terms of fit. A somewhat different interpretation obtains if we use the standard deviation of residuals. For the first model, $s = 0.20$ while for the second model, $s = 0.11$. Thus, a reduction in the standard deviation by 50 percent corresponds to what appears to be a minor improvement in R^2 from 0.95 to 0.98. A comparison of these two measures suggests that the standard deviation of residuals may be more informative of the magnitude of a difference in fit than is true for R^2. However, we suggest that neither measure is appropriate for deciding which of the two models to choose. On the other hand, if we do want to rely on a general measure of fit, the standard deviation of residuals is the more appropriate and useful measure. We explore this difference in the absolute (standard deviation) and relative (R^2) measures of fit in more detail below. In particular, we show several examples in which the value of R^2 can be manipulated while the standard deviation remains the same.

R^2 Manipulations

We assume the simple linear model

$$Y_i = \alpha + \beta X_i + u_i$$

According to this model, the effect of the predictor variable is the same regardless of the value for this variable. If the usual assumptions hold, then $\Sigma \hat{u}_i^2 / (n - 2)$ is an unbiased estimate of σ^2, the variance of the error term. As long as the error variance is constant, this ratio does not change systematically as the set of available observations changes (i.e., $\mathrm{var} u_i$ is constant for all observations). For the same reason, $\Sigma \hat{u}_i^2$ does not change systematically, if the sample size remains constant as the set of available observations changes.

Now, consider the formula for R^2 again:

$$R^2 = 1 - \frac{\Sigma \hat{u}_i^2}{\Sigma (Y_i - \bar{Y})^2}$$

This formula shows that R^2 goes up if the unexplained variation becomes a *smaller proportion* of the total amount of variation in the criterion variable. For example, assume that $\Sigma \hat{u}_i^2 = 10$ for a sample size of five. If the amount of variation in the criterion variable $\Sigma (Y_i - \bar{Y})^2 = 20$, then $R^2 = 1 - 10/20 = 0.50$. On the other hand, if $\Sigma (Y_i - \bar{Y})^2 = 50$, then $R^2 = 1 - 10/50 = 0.80$. Thus, for a given amount of unexplained variation, the higher the total amount of variation in the criterion variable, the higher the R^2 value.

Example

Suppose we are interested in the relationship between savings and disposable income as a simple application of macroeconomics. For the United States, annual data are available indicating total United States savings and disposable income on an annual basis. Based on data for the period 1947–64 the following result is obtained:

$$\hat{S}_t = -2.72 + 0.09 I_t$$

The slope coefficient indicates that as disposable income increases by a dollar (or a billion dollars) savings are expected to increase by nine cents (or 90 million dollars). We also observe the following summary statistics: $R^2 = 0.666$ and $s = 3.40$.

Instead of using savings as a criterion variable, we could have used consumption. From a substantive viewpoint, we should be indifferent about choosing between these alternative criterion variables, because of the equality

$$C_t + S_t = I_t$$

Indeed, based on this equality we can derive the regression result when C_t is the criterion variable. That is,

$$C_t = I_t - S_t$$
$$\hat{C}_t = I_t - \hat{S}_t$$
$$= I_t - (-2.72 + 0.09I_t)$$
$$\hat{C}_t = 2.72 + 0.91I_t$$

Thus, of every additional dollar in disposable income, 91 cents is used for consumption, given that nine cents is used for savings. If we estimate this equation directly, we get exactly the coefficients shown, and the following summary statistics: $R^2 = 0.995$ and $s = 3.40$.

A comparison of the summary statistics across the two equations shows equality of the standard deviation of residuals but inequality in the R^2 values. We now show why the standard deviation of the residuals is the same. Starting with the savings equation

$$\hat{S}_t = -2.72 + 0.09I_t$$

and given that $\hat{u}_t = S_t - \hat{S}_t$, we have

$$S_t = \hat{S}_t + \hat{u}_t$$
$$S_t = -2.72 + 0.09I_t + \hat{u}_t$$

Also,

$$C_t = I_t - S_t$$
$$= I_t - (-2.72 + 0.09I_t + \hat{u}_t)$$
$$C_t = 2.72 + 0.91I_t - \hat{u}_t$$

By comparing the final equations for S_t and C_t we note that the residuals are identical except for the sign. Thus, for a given year the two equations will have residuals of equal value but with opposite signs. As a result, the quantity $\Sigma \hat{u}_i^2$ is identical for the two equations.

We conclude from this that the R^2 value is greater for the consumption equation because of the higher amount of variation in the criterion variable. That is,

$$\Sigma(C_t - \bar{C}_t)^2 > \Sigma(S_t - \bar{S}_t)^2$$

Intuitively it makes sense that the amount of variation in consumption is greater than the amount of variation in savings. For example, savings is a relatively small part of income, while a large part of income is used for consumption. As Lovell says, "the discrepancy between the two [R^2 values] arises because consumption has a bigger variance than savings."[1]

We show the difference between the two equations in Figures 8.1 and 8.2. In Figure 8.1, the slope that represents the estimated effect of income on consumption is very steep. And the variability in the data on the vertical axis (C_t) is substantial. On the other hand, in Figure 8.2, we see a very shallow slope and a small amount of variability along the vertical axis (S_t). Note that in both graphs the amount of variability along the horizontal axis is the same (I_t in both cases), and the residual variation is identical as well.

It is important to keep in mind that one summary statistic, the standard deviation of residuals, is not affected by this manipulation. That is, the absolute amount of unexplained variation does not depend on whether the savings or the consumption equation is chosen. This absolute index is, therefore, a stable indicator. We see later that this absolute index is particularly relevant in other situations where R^2 is an even more ambiguous measure. Before we do that, we want to explore other implications of this apparent anomaly.

Figure 8.1
Consumption as a function of income

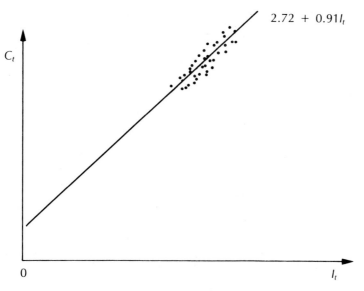

Figure 8.2
Savings as a function of income

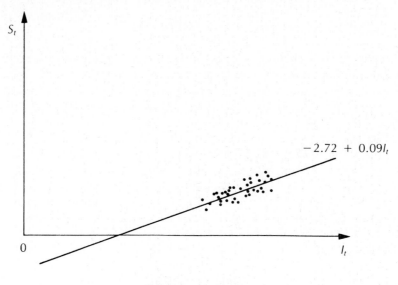

Statistical Tests

We have seen earlier that a test of the null hypothesis H_0: $\beta_1 = \beta_2 = \cdots = \beta_k = 0$ for a multiple regression is based on the unadjusted R^2 value for the equation. For a model with one predictor variable, the F-test is

$$F = \frac{R^2/1}{(1 - R^2)/(n - 2)}$$

Using the savings equation, we obtain $F = 31.9$, while using the consumption equation, we have $F = 3317$. The critical value is $F_{1,16} = 4.49$, at the five-percent level of significance. Thus, we can reject the null hypothesis in both cases, but with substantially more confidence for the consumption equation. In other cases (for example, a smaller sample size), it is possible that the null hypothesis can be rejected with the consumption equation but not with the savings equation. Thus, the F-test is based on the relative explanatory power of the model. In the savings equation we determine the explanatory power of income relative to savings, while in the consumption equation we estimate the explanatory power of income relative to consumption.

Another perspective on this difference can be obtained by comparing the t-tests. Given that the F-test with one degree of freedom in the numerator is the square of the t-test, t-values must be 5.65 and 57.6 respectively. For the savings equation we have

$$t = \frac{0.09}{0.016} = 5.65$$

Similarly, for the consumption equation the t-ratio value is

$$t = \frac{0.91}{0.016} = 57.6$$

We note that the estimated standard error of the slope is the same in the two cases. This is because both the standard deviation of the residuals and the amount of variation in the predictor variable are identical in the two cases. But the magnitudes of the slope coefficients are very different. Thus, the difference in the t-ratios is due to the difference in the slope coefficients. In fact, we can make an argument for the following test procedure. If we are interested in testing the null hypothesis

$$H_0: \beta = 0$$

in the savings equation, then we can choose as an alternative, and equivalent, null hypothesis

$$H_0: \beta = 1$$

in the consumption equation. This follows directly from the equality $C_t + S_t = I_t$, or the fact that every additional dollar not spent on savings ($\beta = 0$ in the savings equation) must be spent on consumption ($\beta = 1$ in the consumption equation). In that case the relevant t-ratio for the consumption equation becomes

$$t = \frac{b - \beta}{s_b} = \frac{0.91 - 1}{0.016}$$
$$= -5.65$$

Thus, by using the dependency between savings and consumption, we are able to obtain equivalent expressions and results from the statistical tests for the slope parameters. (Note that we could have reversed the procedure, and focused on the null hypothesis that $\beta = 0$ in the consumption equation.)

We have shown that both the R^2 value and the results of statistical tests can be influenced by the amount of variation in the criterion variable. Statistical significance is usually required before we are willing to make decisions based on results obtained. However, statistical significance by itself is not an indication that the results are substantively or managerially relevant. Similarly, we often prefer large R^2 values over small R^2 values. But the magnitude of R^2 by itself is not a proper basis for making managerial decisions. And, when the criterion variable differs between two equations, either in terms of its definition or in terms of the data, R^2 values are not comparable. We now turn to such an example.

R^2 for Pooled Data

We have used an incremental R^2 test for the null hypothesis that a subset of the slope parameters in a model is zero (see Chapters 5 and 6). Intuitively, this test determines whether the explanatory power of the full model is significantly better than the explanatory power of a restricted model. (The restriction is that some predictor variables are excluded from the restricted model and have zero influence.) This test suggests that we can compare the R^2 values for different equations. However, the comparison is meaningful only if the criterion variable is *identical* in the different equations. That is, the number and identity of the available observations has to be the same for the comparison of R^2 values to be meaningful. It is, however, tempting to compare R^2 values in other cases as well. We provide an example where some comparisons are meaningful and other comparisons are not.

The Buschmuller Company is interested in estimating a demand equation for PRIMO, its premium beer. Historical data have not been useful for this, however, because the price of PRIMO has remained constant relative to competitors' prices. To assess the impact of the relative price of PRIMO on market share, the firm decided to carry out an experiment. In 1986 six sales territories were selected for this experiment. These territories were considered equivalent in terms of competitors' presence and activity, consumer and retailer characteristics, and so forth.

During the experiment, the relative price (price relative to competitors' prices) of PRIMO was varied across the six territories. Price and other activities of the competitors were monitored closely, but there was no evidence of any systematic differences across the sales territories. The territories were believed to be equivalent in all other respects. At the end of the experiment the following data were obtained:

Territory	Market share[1] of PRIMO in 1986	Relative price[2] of PRIMO in 1986
1	11.5	1.12
2	10	1.11
3	10.5	1.10
4	13	1.09
5	11.5	1.08
6	14	1.07

[1] Unit sales for PRIMO relative to total unit sales for all premium beers.
[2] Price for a six-pack of PRIMO relative to average six-pack price for premium beers.

For the purpose of our illustration, we assume a simple linear model. A regression analysis of market share as a function of relative price (see Table 8.1) showed a slope coefficient of -55.7. Thus, for a 0.01 decrease in the relative price, market share is expected to increase by approximately half a point. However, the statistical significance of the slope coefficient was in doubt, and the R^2 value was very low. For these reasons, the experiment was continued in six different territories. Management decided to use still lower relative prices for these territories. The data at the end of the experiment were as follows:

Territory	Market share of PRIMO in 1987	Relative price of PRIMO in 1987
7	20	1.06
8	18.5	1.05
9	21	1.04
10	22.5	1.03
11	23	1.02
12	25.5	1.01

Based on a separate analysis of the 1987 data, the slope coefficient was found to be -121.4. Thus, in 1987 the expected increase in market share from a 0.01 decrease in relative price is more than one point. Management recognized the possibility that the competitive conditions might not be the same for the two years. Also, as the relative price approaches 1.00 it may be that the price response becomes stronger, suggesting nonlinearity in the relationship between market share and relative price. Nevertheless, management decided that all data should be analyzed together as well. All regression results are shown in Table 8.1.

Inspection of these results makes it clear that the R^2 value for the 1987 data is greater than the R^2 value for 1986. However, the highest R^2 value is obtained when all data are pooled and one equation based on the twelve territories is obtained. The fact that the single equation for the pooled data has a higher R^2 value than for either separate equation suggests to management that this equation is superior. In addition, the vice president of marketing at Buschmuller has indicated that "regression results are of no interest, unless the R^2 value is at least 90 percent." On the other hand, the slope coefficient for the pooled data is -147.6, suggesting an expected increase in market share of 1.5 points for a 0.01 decrease in relative price. This slope coefficient is higher than the value obtained for either the 1986 or 1987 data analyzed separately.

An important question is whether we have in fact a better equation when the data are pooled. Although the R^2 value is higher for the pooled data than for either year analyzed separately, the standard deviation of residuals is worse for the pooled data. In fact, we have argued that R^2 comparisons are meaningful only if the number and identity of the available observations for the different equations is the same. Based on this argument we cannot compare R^2 values across the three equations.

To compare the separate models for 1986 and 1987 with the pooled model, we need to obtain the R^2 value for the pooled data without forcing the parameters to be equal for

Table 8.1
Results for PRIMO

```
                  Regression analysis based on 1986 data

    > COEF

    VARIABLE            B           STD. ERROR(B)          T

    PRICE          -.5571E+02        2.9137E+01         -1.912
    CONSTANT        7.2757E+01        3.1909E+01          2.280

                   MULTIPLE R         R-SQUARE

    UNADJUSTED       0.6911           0.4776
    ADJUSTED         0.5890           0.3469

    STD. DEV. OF RESIDUALS = 1.2189


                  Regression analysis based on 1987 data

    > COEF

    VARIABLE            B           STD. ERROR(B)          T

    PRICE          -.1214E+03        2.5555E+01         -4.752
    CONSTANT        1.4743E-02        2.6453E+01          5.573

                   MULTIPLE R         R-SQUARE

    UNADJUSTED       0.9217           0.8495
    ADJUSTED         0.9010           0.8119

    STD. DEV. OF RESIDUALS = 1.06905


               Regression analysis based on 1986 and 1987 data

    > COEF

    VARIABLE            B           STD. ERROR(B)          T

    PRICE          -.1476E+03        1.4584E+01        -10.118
    CONSTANT        1.7389E+02        1.5540E+01         11.190

                   MULTIPLE R         R-SQUARE

    UNADJUSTED       0.9545           0.9110
    ADJUSTED         0.9498           0.9021

    STD. DEV. OF RESIDUALS = 1.74397
```

the two years. This is accomplished with the following model:

$$Y_{ij} = \alpha + \beta_1 P_{ij} + \beta_2 D_i + \beta_3(P_{ij} * D_i) + u_i$$

where Y_{ij} is the market share for PRIMO in territory i and year j

P_{ij} is the relative price of PRIMO in territory i and year j

$D_i = 1$ if territory i is in year 1987

$\quad\ = 0$ otherwise

This model allows both intercepts and slope coefficients to be different. Given the availability of the separate regression analyses we know that the results will be

$$a = \quad 72.8$$

$$b_1 = -55.7$$

$$b_2 = \quad 74.7$$

$$b_3 = -65.7$$

The unadjusted R^2 for this model allows us to perform an incremental R^2 test.

The full model has an unadjusted R^2 value of 0.9692 and a standard deviation of residuals equal to 1.14642. We can now determine whether the full model is statistically significantly better than the restricted model. The following F-test is used to test the null hypothesis

$$H_0: \beta_2 = \beta_3 = 0$$

$$F = \frac{(R_f^2 - R_r^2)/(DF_r - DF_f)}{(1 - R_f^2)/DF_f}$$

where R_f^2 is the unadjusted R^2 for the full model

R_r^2 is the unadjusted R^2 for the restricted model

DF_f is the degrees of freedom left for the full model

DF_r is the degrees of freedom left from the restricted model

Using the results reported in Table 8.1, and the results for the full model reported above, we have

$$F = \frac{(0.9692 - 0.9110)/2}{(1 - 0.9692)/8} = 7.56$$

The computed F-value compares to a tabulated F-value (five-percent level of significance, 2 and 8 degrees of freedom) of 4.46. We conclude that the full model is superior to the restricted model, at a risk of five percent of being wrong. At least one of the two parameters is different from zero. Thus, it is not appropriate to use the restricted model for the 1986 and 1987 data.

To summarize, the R^2 value for the restricted model reported in Table 8.1 cannot be used as a basis of comparison against the R^2 values for the separate equations also shown in Table 8.1. Only if the criterion variable has the same number and identity of available observations is it appropriate to compare R^2 values. This is the case for the comparison of the full and restricted models. Note, however, that the full model provides exactly the same slope coefficients as the two separate models. Why, then, is the R^2 value for the full model so much higher than the R^2 values for either the 1986 or 1987 data analyzed separately?

To answer this question it is helpful to examine the R^2 formula again:

$$R^2 = 1 - \frac{\Sigma \hat{u}_i^2}{\Sigma(Y_i - \bar{Y})^2}$$

We show the values for the numerator and denominator of the R^2 formula for each of the four models in Table 8.2. A comparison of the two separate models with the full model suggests that the residual sum of squares for the full model is simply the sum of the residual sum of squares for the separate models. However, the sum of squares for the criterion variable in the full model is considerably more than the sum of the same quantities for the separate models. Thus, the full model provides no gain based on unexplained variation. But, an illusion of superiority exists because of the much greater amount of variation in the criterion variable. The magnitude of the increase in this variation depends on the difference in the sample means for the criterion variable in the two data sets. In 1986 the average value for Y is 11.75, while in 1987 it is 21.75. An inspection of the data, shown in Figure 8.3, confirms this.

For the restricted model, the unexplained variation is considerably more than the sum of unexplained variation for the separate models. But the amount of variation in the criterion variable has increased far more, relative to the separate models. Thus, the R^2 for the restricted model makes this model appear very attractive. Finally, we note that

Table 8.2
A comparison of models

	$\Sigma(Y_i - \bar{Y})^2$	$\Sigma \hat{u}_i^2$
Separate model for 1986	11.375	5.9423
Separate model for 1987	30.375	4.5720
Full model for 1986 and 1987	341.750	10.5143
Restricted model for 1986 and 1987	341.750	30.4143

the standard deviation of residuals is not affected by these issues. The full model has a standard deviation of 1.14642, which is the average of the residual standard deviations obtained for the separate models. Of course, the restricted model has a substantially higher standard deviation of 1.74397.

Figure 8.3
A scatterplot of market share versus relative price

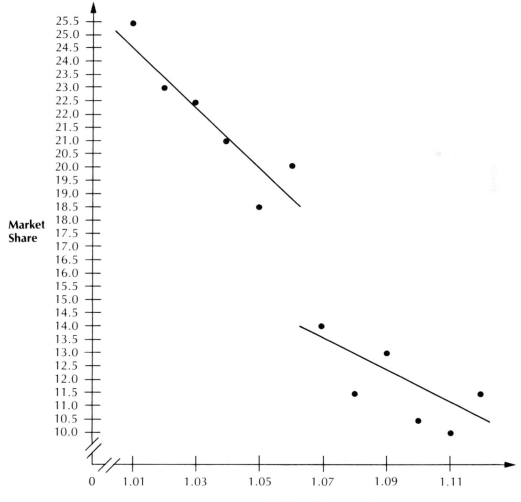

To make this discussion complete, consider the availability of the following data for 1985:

Territory	Market share of PRIMO in 1985	Relative price of PRIMO in 1985
13	10	1.12
14	8.5	1.11
15	11	1.10
16	12.5	1.09
17	13	1.08
18	15.5	1.07

Based on a regression analysis of these data, we obtain a slope coefficient of -121.4 and R^2 equal to 0.8495 (identical to the result for 1987). What is the R^2 value if the 1985 and 1986 are analyzed together?

To answer this question, we focus on some of the same statistics as presented in Table 8.3. What is different about the pooled data for 1985 and 1986 is that the mean value of market share is identical in the two years. For that reason, the total amount of variation in Y for the pooled data is simply the sum of the variation for the separate years. And, in the full model where the intercepts and slope coefficients are allowed to be distinct, the sum of the squared residuals is the addition of the separate values for the two years. Thus, the R^2 value for the full model is now

$$R^2_{\text{full model}} = 1 - \frac{10.5143}{41.75} = 0.748$$

That is, it is between the values for the separate models (although it is not halfway between the separate values).

Aggregation

Aggregation (over time, over individual units) can create specification errors in the model and difficulties in interpreting the results (see Chapter 9). Aggregation also influences the observed R^2 value and the statistical inference (e.g., t-ratios). However, the R^2 value for a model based on aggregate data can be systematically higher or lower

Table 8.3
A comparison of models

	$\Sigma(Y_i - \bar{Y})^2$	$\Sigma\hat{u}_i^2$
Separate model for 1985	30.375	4.5720
Separate model for 1986	11.375	5.9423
Full model for 1985 and 1986	41.750	10.5143

than the value obtained for the equivalent model for disaggregate data.[2] In one study on bank adjustments to monetary policy, an analysis based on weekly data produced $R^2 = 0.52$ ($\bar{R}^2 = 0.50$), while the same model using monthly data provided $R^2 = 0.83$ ($\bar{R}^2 = 0.80$). The conclusions about bank behavior also differed dramatically between the two results.

Removing Cross-Sectional Variation

In the previous example, we used cross-sectional variation to learn about price effects. This is appropriate if the cross-sectional observations (territories) are similar in all relevant aspects (e.g., identity and number of competitors). We concluded, at least for the 1986 and 1987 data, that the parameters should not be constrained to be the same for the two years. Thus, there are either systematic differences between the two years, or differences between the two sets of sales territories, that need to be considered. Also, the linearity assumption is doubtful (see Chapter 6).

We now consider another example, in which there is both cross-sectional and time-series variation. Data are available on a brand's sales, as well as price and other variables, for a large number of stores over many weeks. The brand's average sales level varies considerably across the stores (e.g., due to the size of the stores). Thus, whatever sales differences exist between the stores is due to the size and characteristics of the shoppers, and not necessarily due to any price differences that may also exist between the stores.

The objective of a study may be to explain sales variation for a brand over time, based on changes in the price and in other variables. If the interest is in establishing a price policy for all stores, it is not useful to do a separate analysis for each store. Instead, we may propose to use the data from all stores at the same time in such a manner that the results depend on the time-series and not on the cross-sectional variation.

For simplicity, we assume again a simple linear model with the brand's price as its only predictor variable. That is[4]

$$S_{i,t} = \alpha + \beta P_{i,t} + u_{i,t}$$

where $S_{i,t}$ is the brand's sales in store i in week t, and

          $P_{i,t}$ is the brand's price in store i in week t.

We now consider two options to remove cross-sectional variation, and we let the price effects be determined by time-series variation. The first option is to add a predictor variable that captures the average sales for the brand in a given store, that is,

$$S_{i,t} = \alpha + \beta_1 \bar{S}_i + \beta_2 P_{i,t} + u_{i,t}$$

where $\bar{S}_i$ is the average sales for the brand in store i.

This model change does not affect the criterion variable, which still has both cross-sectional and time series variation. However, the model now will explain the average sales differences for the brand between stores with the additional predictor variable. The price variable should then explain the variation in the brand's sales over time.

There are several possible objections to this option. For example, to the extent that there are average price differences between the stores, the two predictor variables will be somewhat correlated. Although the magnitude of this correlation may not be high (i.e., we do not need to be concerned about multicollinearity), the existence of correlation makes it possible for the price variable to explain some cross-sectional variation in the brand's sales. In other words, the value for β_1 is not necessarily equal to 1. Thus, the interpretation of the price coefficient (the estimate of β_2) cannot be made purely in terms of time-series variation.

A second objection may be made regarding the R^2 value and the F-statistic. To determine the statistical significance of the model, we test the null hypothesis H_0: $\beta_1 = \beta_2 = 0$. If this null hypothesis is rejected, we know that at least one of the parameters is not zero. But the first parameter cannot be zero (unless the average sales values across the stores are exactly equal, in which case there is no variation in $\bar{S}_i$). The explanatory power provided by the first predictor variable is artificial and has no relevance to the purpose of the study. Thus, the R^2 value is inflated, in the sense that it incorporates the explanatory power provided by the average store sales values.

The second option is to adjust both sales and price variables by their average values. For example, we can subtract the average store sales and average store price for the brand from the two variables:

$$(S_{i,t} - \bar{S}_i) = \alpha + \beta(P_{i,t} - \bar{P}_i) + u_{i,t}$$

where $\bar{P}_i$ is the average price for the brand in store i.

With this formulation all cross-sectional variation is removed, so that the parameter estimates are based only on the time-series variation. In addition, the R^2 value (and hence F or t statistics) captures only the power of the model for explaining time series variation. Of course, the R^2 value for the second option is not comparable to the R^2 value for the first option, because the definitions of the criterion variables are not the same. Yet it is relevant, given a tendency to concentrate on R^2 values, to know that the second option will *necessarily* result in a lower R^2 value than the first option. The reason for this is that the first model has more variation, and the additional variation is explained by an additional predictor variable. Thus, the use of R^2 as an index of a model's quality would lead us to choose the first option, even though the second option is preferable, if the objective of the study is to exploit time-series variation.

Removing Irrelevant Variation

Another illustration of an artificial increase in the R^2 value is contained in the following example.[5] The authors of the study were interested in developing a model to relate managerial compensation to characteristics of MBA applicants known prior to admission to the MBA program. The criterion variable was defined as the annual compensation of a graduate several years after leaving the program. Predictor variables included the GMAT scores, undergraduate grade-point average, undergraduate major, and so forth. Because the number of observations available for a given graduating class was small, data were obtained from graduates representing seven consecutive years. To a substantial extent, current compensation is influenced by the number of years since graduation.

To adjust for this difference between the graduating classes, the authors used two different procedures. The first model includes "Years of work experience after MBA" and "Years of work experience prior to MBA" as separate predictor variables (in addition to GMAT, undergraduate major, etc.). The inclusion of the experience variables allows the model to account for average differences in current compensation between the graduating classes. The second procedure consists of two regression analyses. Current compensation is explained as a function of *only* the work-experience variables. This equation is used to create a new compensation variable that is adjusted for differences in the number of years of work experience. Then the adjusted compensation variable is regressed against GMAT, undergraduate major, and so forth. For (staff) managerial compensation, the authors report a multiple R of 0.75 ($R^2 = 0.56$) for the first model and a multiple R of 0.38 ($R^2 = 0.14$) for the second model (based on adjusted compensation).

If the objective of the study is to relate characteristics of MBA applicants to subsequent annual compensation, then the fit of the first model overstates the case. Thus, if the focus is on the explanatory power of variables such as GMAT score, then the second model is more appropriate because differences in compensation due to differences in work experience are removed. The second model is preferred because the irrelevant variation is removed. This is accomplished by removing the variation in the criterion variable that is accounted for by an otherwise irrelevant variable. The second model may approximate the results for a large sample of observations from one graduating class.

Interestingly, the standard deviation of residuals is the same in the two cases. That is, whether one includes years of work experience as a predictor variable to explain compensation variation between the graduating classes (as in the first procedure), or adjusts compensation for differences in work experience (as in the second procedure), the *absolute* amount of unexplained variation is the same.

Shifting Validities

In some areas of inquiry, such as in psychology, there is a strong tendency to rely on correlation coefficients for the determination of the validity of a regression equation. For example, suppose we are interested in describing the validity of the GMAT score of applicants to an MBA program for predicting academic performance. The higher this predictive validity, the more admissions officers may want to rely on the GMAT score for determining the ability of an applicant to satisfy the requirements of the MBA program.

When the GMAT was first developed (the acronym was then ATGSB), no information was available on its predictive validity. To determine the usefulness of the test scores, several studies were done to relate the academic performance of MBA students to their GMAT score. The sample consisted of students who had taken the test at a time when the test results were not yet used for the selection of applicants. Based on these data, correlation coefficients were computed to estimate the linear association between the GMAT score and the first-year grade-point average in an MBA program. For one school this validity coefficient was 0.64. Partly based on this result, the school's administration became convinced that the GMAT score would be a useful measure for predicting academic performance. Five years later, however, when the GMAT scores had become a regular part of the selection procedures at the school, the correlation coefficient was found to be only 0.44.

To understand this anomaly, we assume that it is appropriate to consider only linear effects, and that there are no other relevant predictor variables. Under these conditions the simple correlation coefficient is an appropriate measure of the degree of association between the two variables. However, the correlation coefficient will be influenced by the amount of variation available in the variables. Given that the admissions committee uses the GMAT score as a basis for admissions decisions, we concentrate on this variable.

When the GMAT is not used as a selection variable, the actual GMAT scores may vary over a large part of the range of possible scores. However, when the GMAT is a selection variable, the range of variation on this variable tends to be reduced. This restriction in the range of variation for the predictor variable is responsible for the decline in the correlation coefficient.

One way to think about this problem follows. Academic performance is influenced by many factors, including a student's innate ability (partly measured by the GMAT score), motivation to perform, distractions, resources, and so on. When students vary considerably in innate ability (e.g., the students vary greatly in their GMAT scores), this factor may have considerable power to explain differences in academic performance. However, when students are selected based on a measure of innate ability (GMAT), and applicants scoring low on this measure are not admitted, the relative explanatory power of the other factors is enhanced. Thus, in the case of one school cited above, when the GMAT scores are not restricted, this variable accounts for about 40 percent of the variation in academic performance ($r = 0.64$). But, when this predictor variable is restricted, its explanatory power may be reduced to about 20 percent ($r = 0.44$). Note that the actual explanatory power of the GMAT may not have changed (i.e., in terms of the total applicant population). The available data, however, are limited to students not rejected based on a low GMAT score. This restriction is responsible for the illusion of a loss in validity.

We can imagine that there are similar implications for other application areas. For example, if participants in a particular sport for the Olympics are largely selected based on certain physical characteristics, variation in the performance of the selected individuals at the Olympics will be difficult to explain with the remaining variation in physical characteristics. Certainly the R^2 value obtained is a function of the remaining heterogeneity in these physical characteristics.

We have provided several diverse illustrations to explain what may influence R^2. These examples should make it clear that the magnitude of R^2 in general is a poor basis for judging the quality or usefulness of a regression result. We have shown how the R^2 value depends on the choice of criterion variable, on the combination of cross-sectional and time-series variation, and on the presence of irrelevant variation that can be accounted for. An alternative measure of overall goodness of fit for the model, the standard deviation of residuals, is not subject to the various manipulations. Thus, if an overall measure is desired, we should examine this standard deviation and not rely on R^2 exclusively. The last example of shifting validities also has implications for the use of standardized slope coefficients, a topic which we discuss below.

Standardized Regression Coefficients

Many software packages for regression analysis include standardized regression coefficients, or beta weights. Often, standardized regression coefficients are used to infer the relative importance of the predictor variables, where relative importance is measured by relative explanatory power. The reason for using these measures is that the slope coefficients themselves are not comparable. For example, suppose that an admissions director wants to compare the influence of the GMAT score to the impact of the undergraduate GPA on the academic performance in graduate school. Given that the GMAT is measured on an 800-point scale and undergraduate GPA is measured on a 4-point scale, the slope coefficients are clearly not comparable. Standardization provides some measure of comparison.

Consider first the simple linear model:

$$Y_i = \alpha + \beta X_i + u_i$$

For this model, the least-squares estimate of β is

$$b = \frac{\sum X_i Y_i - \dfrac{\sum X_i \sum Y_i}{n}}{\sum X_i^2 - \dfrac{(\sum X_i)^2}{n}}$$

Also, a measure of the linear association between Y and X is obtained with the correlation coefficient

$$r = \frac{\sum X_i Y_i - \dfrac{\sum X_i \sum Y_i}{n}}{\sqrt{\sum X_i^2 - \dfrac{(\sum X_i)^2}{n}} \ \sqrt{\sum Y_i^2 - \dfrac{(\sum Y_i)^2}{n}}}$$

Note that the numerators in the formulas for b and r are identical. We have also shown in Appendix 2.6 that

$$r = b \frac{s_x}{s_y}$$

Now, if we define

$$Y_i^* = \frac{Y_i - \bar{Y}}{s_Y}$$

$$X_i^* = \frac{X_i - \bar{X}}{s_X}$$

and do a regression analysis based on Y^* and X^*, the slope coefficient for this regression analysis is r. Thus, we can interpret the correlation coefficient as a standardized slope coefficient. The correlation coefficient measures the estimated number of standard deviation units by which Y changes if X is increased by one standard deviation unit.

Multiple Regression

Based on a comparison of correlation and slope coefficients in a simple linear regression analysis, it is easy to see the appeal of a similar standardization for multiple regression analysis. For the model

$$\hat{Y}_i = a + b_1 X_{1i} + b_2 X_{2i} + \cdots + b_m X_{mi}$$

we define

$$\text{beta}_1 = b_1 \frac{s_{X_1}}{s_Y}$$

$$\text{beta}_2 = b_2 \frac{s_{X_2}}{s_Y}$$

$$\vdots$$

$$\text{beta}_m = b_m \frac{s_{X_m}}{s_Y}$$

These beta weights are comparable in the sense that they are not affected by the unit of measurement for a predictor variable. For example, if X_1 is measured in thousands of dollars, b_1 is an estimate of the effect of a \$1,000 increase in X_1, holding the other predictor variables constant. However, if we measure X_1 instead in dollars, b_1 is an estimate of the effect of a \$1 increase in X_1, holding the other predictor variables constant. In the second case the estimated slope coefficient will be 1,000 times larger[6]. On the other hand, the beta weight is not affected by such a change in measurement, because the standard deviation of X_1 will be 1,000 times smaller in the second case. Thus, the beta weight measures the expected standard-deviation unit change in Y for a one-unit standard-deviation increase in a given predictor variable, holding the other predictor variables constant.

Uncorrelated Predictor Variables

If all of the predictor variables are uncorrelated in terms of pairs, then the slope coefficients in a multiple regression analysis are equal to the corresponding slope coefficients in separate simple regression analyses. It follows that in this case the beta weights equal the simple correlation coefficients between the criterion variable and each predictor variable. And the sum of the squared beta weights (or the squared correlation coefficients) equals the R^2 value for the equation (see Appendix 4.2). Thus, in the case of uncorrelated predictor variables, the beta weights have a very straightforward interpretation. It is easy to see that the beta weights provide a measure of relative explanatory power for the predictor variables.

Correlated Predictor Variables

Usually we encounter data sets with correlated predictor variables. Now, the existence of collinearity influences the range of possible values for beta weights. For example, beta weights are contained in the $[-1, +1]$ interval when the predictor variables are uncorrelated. Collinearity, however, makes it possible for the beta weights to fall outside this interval. Thus, if beta weights occur outside this interval, there is reason to consider the issue of multicollinearity. The interpretation of the beta weights is also more difficult in such a case. The higher the estimated standard error associated with a slope coefficient, the greater the range of possible values for the beta weight. Thus, the predictor variable that has the highest degree of multicollinearity (see Chapter 5) is also likely to have the highest (absolute) beta weight. Consequently, multicollinearity confounds the interpretation of beta weights in terms of explanatory power. There are, however, procedures developed specifically for obtaining reliable importance weights under conditions of multicollinearity[7]. We now turn to other difficulties related to the use of beta weights.

Factors Influencing Beta Weights

We can express the criterion variable in a simple linear model as

$$Y_i = \alpha + \beta X_i + u_i$$

Also, assuming that u_i and X_i are independent (see Chapter 9), we have

$$\text{Var}(Y_i) = \beta^2 \, \text{var}(X_i) + \text{var}(u_i)$$

Now, R^2 is defined as the ratio of explained variation over total variation in Y, or

$$R^2 = \frac{\beta^2 \ \text{var}(X_i)}{\beta^2 \ \text{var}(X_i) + \text{var}(u_i)}$$

With one predictor variable $R^2 = r^2 = \text{beta}^2$. Thus, based on this R^2 formula, the beta weight is influenced by three factors: (1) the magnitude of the slope parameter, β, (2) the variance in the predictor variable, and (3) the variance of the residuals. It is clear that, holding β and $\text{var}(u_i)$ constant, the larger the variance of the predictor variable in the sample, the larger the beta weight.

If instead our model contains two predictor variables, we have

$$Y_i = \alpha + \beta_1 X_{1i} + \beta_2 X_{2i} + u_i$$

Assuming that the predictor variables are uncorrelated, we have

$$\text{Var}(Y_i) = \beta_1^2 \ \text{Var}(X_{1i}) + \beta_2^2 \ \text{Var}(X_{2i}) + \text{Var}(u_i)$$

and the squared beta weights are

$$\text{beta}_1^2 = \frac{\beta_1^2 \ \text{Var}(X_{1i})}{\beta_1^2 \ \text{Var}(X_{1i}) + \beta_2^2 \ \text{Var}(X_{2i}) + \text{Var}(u_i)}$$

$$\text{beta}_2^2 = \frac{\beta_2^2 \ \text{Var}(X_{2i})}{\beta_1^2 \ \text{Var}(X_{1i}) + \beta_2^2 \ \text{Var}(X_{2i}) + \text{Var}(u_i)}$$

Furthermore, the ratio of the squared beta weights is

$$\frac{\text{beta}_1^2}{\text{beta}_2^2} = \frac{\beta_1^2 \ \text{Var}(X_{1i})}{\beta_2^2 \ \text{Var}(X_{2i})}$$

The ratio of the squared beta weights makes it clear that beta_1^2 will be greater than beta_2^2, if $\text{Var}(X_{1i}) > \beta_2^2/\beta_1^2 \ \text{Var}(X_{2i})$.

If we require that beta weights are *stable* measures, both the slope parameters and the sample variances of the predictor variables must be stable. For example, consider a model for the demand of a product as a function of its price and advertising expenditures. What is the meaning of the sample variances for these predictor variables? Assuming that both price and advertising are determined by management, the sample variances are a function of market and other characteristics used by management to set the levels of these variables. If management varies price considerably but maintains advertising at a relatively constant level, it is unlikely that advertising will have the higher beta weight. In other words, the sample variances of price and advertising are not stable measures.

To dramatize the problem, consider two managers who compete with identical products in identical markets. Both managers can vary price and advertising for the product. Suppose that the first manager manipulates price a great deal but keeps advertising at a fairly constant level. The second manager varies advertising considerably and holds the price level approximately constant. A multiple regression analysis for the first manager's sales data shows that price has the largest squared beta weight. For the second manager's sales data, advertising has the largest squared beta weight. For the first manager the results confirm that price is the most important variable, while for the second manager the same analysis confirms the greater importance of advertising.

Instead of relying on beta weights, a manager should study profit consequences for the determination of price and advertising decisions. Indeed, for predictor variables which are subject to manipulation (such as price and advertising) beta weights are meaningless and should not be computed. On the other hand, if there is a *natural variance* for each predictor variable, then it is appropriate to use beta weights. For example, if the observations describe characteristics of households such as political attitudes, amount of television watching, and so on, it is possible to use beta weights. However, we must then *define* the population about which we claim to have some knowledge and use probability sampling for data collection. Only in that case is the sample variance for each predictor variable an unbiased estimate of the variance in the population, and in such a case it is meaningful to use beta weights.

Probability sampling is also critical to the use of beta weights in a study of the relationship between attitudinal measures and other items defined on somewhat arbitrary scales. Of course, the slope coefficients are difficult to interpret in such a case. For example, the value of the slope coefficient would depend on whether an attitude is measured on a five-point or an eleven-point scale. For that reason, survey research results based on regression analysis often are stated in terms of beta weights. However, probability sampling of the population of interest is required for generalizability of the results.

SUMMARY

In this chapter we have emphasized the need to be very careful about interpreting R^2 values. Regression analyses are commonly summarized in terms of the model's explanatory power. It is often said that the higher the R^2 value, the better the model. However, we have shown several examples where reliance on R^2 may lead to wrong model choices. If it is important to use an overall measure, then the standard deviation of residuals should be preferred.

The examples we have used vary somewhat in terms of the difficulties involved in the use of R^2. The AT&T dividend example is meant to suggest that a model that provides understanding (dividends as a function of earnings) should be preferred over a model that does not, even the R^2 value is lower for the first model. The simple application of macroeconomics (consumption and savings) shows how the choice of criterion variable can influence R^2, even though the alternatives provide identical information.

The third example shows how a model based on both cross-sectional and time-series variation may have a higher R^2 value than separate models based on only one source of variation. We showed the source of increase in R^2, and demonstrated that a choice based on R^2 values may be wrong. Use of the standard deviation of residuals avoids the problem. Aggregation also influences the R^2 value observed. However, the effect of aggregation (over time periods or across individuals) depends on characteristics of the data. The R^2 value obtained from aggregate data can be higher or lower than the value obtained from an equivalent model using disaggregate data. Either way, however, an analysis of disaggregate data is more likely to provide valid results.

Additional manipulations involve how adjustments can be made for "irrelevant" variation. If we have two sources of variation in the data, and we are interested in using only one source, we can add a predictor variable to account for the "irrelevant" variation. Or, we can adjust the criterion variable directly. We prefer to use a regression with the adjusted criterion variable, even though it produces a lower R^2 than the other procedures. Again, there is no difference in the standard deviation of residuals between the two procedures.

The final example shows how restricted variation in a predictor variable influences the correlation coefficient (and hence R^2). That is, the more a predictor variable is used to select observations (e.g. individuals) for a program, the lower the potential explanatory power of that variable. Thus, the range of sample variation in a predictor variable is an important determinant of the R^2 value. This notion also is the primary basis for our discussion of beta weights.

When predictor variables are measured in noncomparable units, it is impossible to compare the impact of these variables based on the slope coefficients. For that reason, slope coefficients are often standardized (to remove the influence of the different measurement scales). However, these measures are influenced by the amount of sample variation present in the predictor variables. The higher the variation in a predictor variable (holding the variation in the other predictors constant), the higher its beta weight. Thus, we should critically evaluate the use of beta weights. Are they really necessary and useful? We propose that beta weights are only valid if there is a natural amount of variation for the predictor variables, and if probability sampling is used to obtain unbiased estimates of the population variance in the predictor variables.

In general, the measure(s) we use to summarize regression results should be relevant to the purpose of the study. If the main purpose is to develop a forecasting model, we should focus primarily on the standard error of forecasts developed, or we could focus on the sensitivity of the forecast to alternative scenarios. If the main purpose is to understand the relationship between variables, we should focus on the precision of the slope coefficient(s), and their plausibility for a range of possible values for the predictor variables. If we do need to use a summary measure, the standard deviation of residuals is to be preferred over R^2.

ENDNOTES

1. Lovell, Michael C. *Macroeconomics: Measurement, Theory, & Policy* (New York: John Wiley & Sons, 1975), p. 145.

2. See, Rowe, Robert D., "The Effects of Aggregation over Time on *t*-Ratios and R^2's," *International Economic Review*, 17:3 (October 1976): 751–7.

3. Bryan, William R., "Bank Adjustments to Monetary Policy: Alternative Estimates of the Lag," *American Economic Review*, 57 (September 1967): 855–864.

4. If the stores differ in size, then the effect of a price change on sales should depend on the size of the store. Such a complication can and should be accommodated either in the measurement of the variables or in the model specification.

5. Weinstein, Alan G., and V. Srinivasan, "Predicting Managerial Success of MBA Graduates," *Journal of Applied Psychology*, 59:2 (1974): 207–12.

6. Assuming, of course, that the criterion variable is the same in terms of its values and its measurement.

7. See, for example, Green, Paul E., J. Douglas Carroll, and Wayne S. DeSarbo, "A New Measure of Predictor Variable Importance in Multiple Regression," *Journal of Marketing Research*, 15 (August 1978): 356–60.

REFERENCES

Bryan, William R. "Bank Adjustments to Monetary Policy: Alternative Estimates of the Lag." *American Economic Review*, 57 (September 1967): 855–864.

Green, Paul E., J. Douglas Carroll, and Wayne S. DeSarbo. "A New Measure of Predictor Variable Importance in Multiple Regression." *Journal of Marketing Research*, 15 (August 1978): 356–60.

Lovell, Michael C. *Macroeconomics: Measurement, Theory, & Policy*. New York: John Wiley & Sons, 1975.

Rowe, Robert D. "The Effects of Aggregation over Time on *t*-Ratios and R^2's." *International Economic Review*, 17:3 (October 1976): 751–7.

Weinstein, Alan G., and V. Srinivasan. "Predicting Managerial Success of MBA Graduates." *Journal of Applied Psychology*, 59:2 (1974): 207–12.

9

Special Topics

In the early chapters of this book, we have made the implicit assumption that predictor variables are fixed. In this chapter we consider some additional issues that may be relevant to regression analysis when this assumption is not tenable. Specifically, we address the problems of *simultaneity, lagged criterion variables,* and *errors in variables.* We also offer some additional discussion of the problems of aggregation and pooling.

The Simple Model Revisited

For the simple regression model

$$Y_i = \alpha + \beta X_i + u_i$$

where

$$i = 1, 2, \ldots, n$$

we can show that the least-squares estimator

$$b = \frac{\sum X_i Y_i - \dfrac{\sum X_i \sum Y_i}{n}}{\sum X_i^2 - \dfrac{(\sum X_i)^2}{n}}$$

is unbiased, if both of these conditions hold:

  a. $E(u_i) = 0$ for all i
  b. X_i is fixed   for all i

The assumption that X_i is fixed can be interpreted as follows. A manager may manipulate or control the predictor variable and choose whatever values he or she wants to. The particular values observed for the predictor variable in a sample would not depend on probability sampling or other random processes. Thus, we assume that the predictor variable is not a random variable.

The simple model indicates that the outcome for the criterion variable depends on the specific value for the predictor. But even if the parameters, α and β, are known, we would not be able to claim exact knowledge about this outcome. That is, the error term, u_i, adds a random element to the relationship of interest. Since Y_i depends on both X_i and u_i, the criterion variable has to be a random variable. And, as we argue in Chapter 3, the key to statistical inference for a regression analysis is the correct specification of the model. By analogy, if we want to test a hypothesis about the mean of a population, we need to use probability sampling to justify the use of test statistics. In regression, we make assumptions about the error term which are equivalent to the implications of probability sampling. Whether these assumptions are appropriate must then be determined (for example through residual analysis). But the use of a well-specified model is a necessary condition for these assumptions to be satisfied (all relevant predictor variables must be included, nonlinearities accommodated, etc.).

Sometimes it is impossible to claim that the values of the predictor variable are fixed. For example, the predictor variable may have a random component (see the discussion of errors in variables below) or it may depend on the criterion variable (see the discussion of simultaneity below). In that case, we cannot assume that the predictor variable is fixed. The question then is whether we can substitute the assumption that X_i and u_i, both being random variables, are independent. This substitute assumption, if valid, would allow us to justify the statistical inference procedures we have applied.

To see the similarity between this substitute assumption and the original one, consider a model that excludes a relevant predictor variable. In that case, the error term reflects the influence as well as the specific value of that predictor variable. Now, if this missing predictor is *uncorrelated* with the included predictor variables, it is reasonable to assume that the error term is independent of the included predictors. However, if the missing predictor is correlated with one or more included predictor variables, this assumption cannot hold. Intuitively, the problem is that the correlation between included and excluded variables allows the included variables to account for more variation in the criterion variable than they should. Thus, parameter estimates will be biased, with the nature and magnitude of the bias dependent upon the correlations and the true effect of the omitted variable. In the following section, we discuss another reason why X_i and u_i may not be independent.

Simultaneity

Suppose there are two variables of interest, Y and X. The relationship between these variables is considered to be *simultaneous*, if Y depends on X *and* X depends on Y. For example, let Y represent sales of a product, and let X be the advertising effort. We are interested in understanding the effect of advertising on sales. For simplicity, assume that sales is a linear function of advertising, and that advertising is also a linear function of sales:

$$Y_i = \alpha + \beta X_i + u_i \tag{1}$$

$$X_i = \gamma Y_i + v_i \tag{2}$$

where Y_i is sales of the product in territory i,

X_i is advertising for the product in territory i,

α, β, and γ are unknown parameters, and

u_i and v_i are error terms.

Both equations have an error term, and therefore both Y and X are random variables. In addition, we have assumed a zero intercept for the advertising equation. This is consistent with the idea that advertising may be determined as a percentage of sales. The presence of an error term in the advertising equation indicates that there is additional (random) variation in the advertising amount between territories that is not accounted for by sales.

Given that X is a random variable, we need independence between X and u to be able to justify the statistical inference procedures using the least-squares estimates. However, if (1) and (2) are true, it is impossible for X and u to be independent. To see this, we substitute (1) into (2):

$$X_i = \gamma[\alpha + \beta X_i + u_i] + v_i$$
$$= \alpha\gamma + \beta\gamma X_i + \gamma u_i + v_i$$
$$X_i = \frac{\alpha\gamma}{1 - \beta\gamma} + \frac{\gamma}{1 - \beta\gamma} u_i + \frac{v_i}{1 - \beta\gamma}$$

It is clear from the result of the substitution that X_i depends on u_i. The strength of this dependency is determined to a large extent by γ, the parameter of the advertising equation. Thus, we conclude that it is inappropriate to use the least-squares procedures to estimate the parameters of (1), the sales equation.

To see the problem at an intuitive level, consider the following. Suppose that the advertising amount is determined as a fixed percentage of sales, that is, let

$$X_i = .05Y_i$$

We assume that there is no error term in this equation and that management is able to maintain this relationship exactly across all observations. This requires not only that management allocates five percent of sales to advertising, but that the actual sales level is known for each territory at the time the advertising expenditures are determined. Now, the existence of this advertising equation has implications for our ability to estimate the sales equation.

Specifically,

$$Y_i = \frac{1}{0.05} X_i = 20X_i$$

Thus, if we ignored the existence of the advertising equation and performed a regression analysis of sales as a function of advertising, we would conclude that each additional unit of advertising increases sales by 20 units. Also, the value of R^2 for this regression would be 1.00. We conclude that, if advertising is completely determined by sales, we simply obtain the inverse of this relationship for the equation we are really interested in. This happens regardless of the true effect of advertising on sales. Unfortunately, there is no way out of this dilemma if advertising is determined in this manner.

Underidentification

In general, no valid information is obtainable about the effect of advertising on sales if equation (2) holds. This is called a problem of underidentification. Underidentification occurs when it is impossible to obtain unique parameter estimates for the variables of interest while we properly consider the simultaneity of the relationship. To see this, we use the following equations:

$$Y_i = \alpha + \beta_1 X_{1i} + \beta_2 X_{2i} + u_i \tag{3}$$

$$X_{1i} = \gamma Y_i + v_i \tag{4}$$

where X_{1i} is advertising for the product,

X_{2i} is the price of the product,
and other variables are as defined previously.

As before, the problem with applying least squares to equation (1) is that X_{1i} and u_i are not independent. One way to resolve this problem is to make some substitutions. Specifically, least squares can be applied to each equation by itself, if the right-hand side of the equation does not contain a variable that appears on the left-hand side of another equation.

Substituting (4) into (3) gives us

$$Y_i = \alpha + \beta_1(\gamma Y_i + v_i) + \beta_2 X_{2i} + u_i$$
$$= \alpha + \beta_1 \gamma Y_i + \beta_1 v_i + \beta_2 X_{2i} + u_i$$

$$Y_i - \beta_1 \gamma Y_i = \alpha + \beta_2 X_{2i} + \beta_1 v_i + u_i$$

$$Y_i = \frac{\alpha}{1 - \beta_1 \gamma} + \frac{\beta_2}{1 - \beta_1 \gamma} X_{2i} + \frac{\beta_1 v_i + u_i}{1 - \beta_1 \gamma} \tag{5}$$

Similarly, substituting (3) into (4) gives us

$$X_{1i} = \gamma(\alpha + \beta_1 X_{1i} + \beta_2 X_{2i} + u_i) + v_i$$
$$= \alpha\gamma + \beta_1 \gamma X_{1i} + \beta_2 \gamma X_{2i} + \gamma u_i + v_i$$

$$X_{1i} - \beta_1\gamma X_{1i} = \alpha\gamma + \beta_2\gamma X_{2i} + \gamma u_i + v_i$$

$$X_{1i} = \frac{\alpha\gamma}{1 - \beta_1\gamma} + \frac{\beta_2\gamma}{1 - \beta_1\gamma} X_{2i} + \frac{\gamma u_i + v_i}{1 - \beta_1\gamma} \tag{6}$$

Both resulting equations, (5) and (6), have X_{2i} as the only predictor variable on the right-hand side. Since X_{2i} is not dependent on either Y_i or X_{1i}, there is no difficulty in making the usual assumptions. Thus, least squares can be used on these two equations separately. After obtaining coefficients for these two equations, we have to solve for (unique) values of the parameters in equations (3) and (4).

Suppose that we have applied least squares to each equation and that we have the following "results":

$$\hat{Y}_i = p_1 + q_1 X_{2i} \tag{7}$$

$$\hat{X}_{1i} = p_2 + q_2 X_{2i} \tag{8}$$

Note that if X_{2i} is in fact an irrelevant variable in equation (3), that is, if $\beta_2 = 0$, then the expected value of both q_1 and q_2 is zero (i.e., both q_1 and q_2 have β_2 in the numerator). But if statistically significant (and valid) estimates are obtained, we can use these to derive estimates for the parameters of interest.

By comparing equations (5) and (6) with (7) and (8), we see that the ratio q_2/q_1 provides an estimate of γ. Alternatively, an estimate of γ can be obtained using the ratio p_2/p_1. However, there is no way to obtain *unique* estimates of β_1 and β_2 from these equations. For that reason, the system of equations is *underidentified* with respect to these two parameters. Thus, even if (3) and (4) are correctly specified, they are of no use in learning about advertising's effect on sales.

The idea we pursued above hinges on removing certain variables from the right-hand side of the original equations. The process of substitution results in two new equations, often referred to as *reduced-form* equations. Under the usual assumptions, least squares can be used separately for each of the reduced-form equations, and if there is enough information, estimates of the parameters in the original equations can be obtained indirectly. This procedure is referred to as *indirect least squares*.

We note that underidentification is avoided if there is in fact another predictor variable relevant to the advertising equation. For example, suppose that advertising is determined by sales and cash flow. We can then hypothesize the following system of original equations:

$$Y_i = \alpha_1 + \beta_1 X_{1i} + \beta_2 X_{2i} + u_i \tag{9}$$

$$X_{1i} = \alpha_2 + \gamma_1 Y_1 + \gamma_2 X_{3i} + v_i \tag{10}$$

where X_{3i} is cash flow and other variables are as defined before.

We have also added an intercept to the advertising equation to give it more flexibility. If all parameters included in (9) and (10) truly differ from zero, then it is possible to obtain unique estimates for all parameters through the method of indirect least squares.

Comments

Complex estimation procedures have been developed for problems involving simultaneous causality (as in the example above, where sales is a function of advertising, and advertising is a function of sales). These procedures include *indirect least squares*, *two-stage least squares*, *three-stage least squares*, and *limited information estimators*. The practice of considering simultaneity for empirical work is both appropriate and important. Results can be very misleading if existing simultaneity is ignored. On the other hand, the effective sample size needed for reliable estimation of parameters is larger if simultaneity is accommodated, relative to the sample size needed for least squares. In addition, the problem of model specification is considerably more complex in the face of simultaneity. Two reasons for this are (1) that we have to recognize all relevant dependencies for the predictor variables in the equation of interest (i.e., the sales equation in our example), and (2) that we have to have correct model specifications for each equation. (E.g., an error in the specification of the advertising equation will affect the validity of the estimates of parameters in the sales equation.)

Under certain conditions, it may be appropriate to ignore dependencies for predictor variables. For example, if the dependency is a weak one, the bias in the least-squares estimator will be small. Another plausible condition has to do with aggregation. Suppose that the advertising amount is determined on an annual basis, and that the actual support (proportion of sales) for advertising fluctuates widely within the year. If the sales equation can be specified for, say, weekly data, the dependency that exists at the annual level may be ignored. Also, if the dependency exists for lagged periods, we cannot claim a bias for the least-squares procedure. For example, suppose advertising is determined as a function of the previous period's sales, that is,

$$X_{1it} = \gamma Y_{i,t-1} + v_t \tag{11}$$

In this case, the existence of an advertising equation does not make it inappropriate to use least squares for the sales equation. Under the usual assumptions required for the parameters in the sales equation to be unbiased, we can apply least squares directly to equation (3). Note, however, that we cannot assume that X_1 is fixed, since the values depend on past sales.

Lagged Criterion Variables

If the sample available for a regression analysis consists of time-series data, the model often includes a *lagged criterion variable*. In this section, we explore some reasons for the use of lagged values of the criterion variable on the right-hand side of the equation. The three cases considered are (1) geometrically declining parameters for lagged effects of a predictor variable (the Koyck specification[1]), (2) partial adjustment of the behavior under study to changes in relevant conditions, such as a deviation from optimal behavior due to ignorance, inertia, and so on, (3) adaptive expectations, where the behavior is determined based on the expected levels of predictor variables. In each of the three cases, certain assumptions are made which ultimately result in an equation containing lagged values of the criterion variable as a predictor variable.

The Koyck Specification

Suppose that we have a model to estimate the effect of a predictor variable, X_t, on the criterion variable, Y_t. However, the effect of X_t is expected to last beyond the current period. That is, let

$$Y_t = \alpha + \beta_1 X_t + \beta_2 X_{t-1} + \beta_3 X_{t-2} + \cdots + \beta_m X_{t-(m-1)} + u_t \qquad \textbf{(12)}$$

Equation (12) allows the lagged effects of the predictor variable to exist for an unlimited period of time. Indeed, the difficulty with (12) is that it does not specify how many lagged terms should be included. Unless m is small, the model is difficult to estimate. For example, suppose $m = 10$. In that case, we have ten predictor variables which contain similar information. Often these predictors are correlated, with the result that we may have great difficulty obtaining reliable parameter estimates for the separate lagged terms. This is especially true if X_t varies systematically (e.g., if there is a positive trend in X_t). In addition, the number of observations available for estimation is reduced by $(m - 1)$. Thus, if $m = 10$, and the original sample size equals 40, the effective sample size would be 31. In practice, the choice of m would be arbitrary.

To avoid these problems, we can impose constraints on the relation between predictor variable (the Koyck specification[1]), (2) partial adjustment of the behavior under study to changes in relevant conditions, such as a deviation from optimal behavior

$$\beta_2 = \lambda \beta_1$$

$$\beta_3 = \lambda \beta_2 = \lambda^2 \beta_1$$

$$\vdots$$

$$\beta_m = \lambda \beta_{m-1} = \cdots = \lambda^{m-1} \beta_1$$

where $\quad 0 \leq \lambda < 1$

If the value of λ is between zero and one, then there are lagged effects for X, and the effects for all lagged periods can be expressed as a fraction of β_1, the current effect. As shown above, this fraction declines geometrically, based on the number of lags.

With these restrictions, we can rewrite (12) as

$$Y_t = \alpha + \beta_1 X_t + \lambda\beta_1 X_{t-1} + \lambda^2\beta_1 X_{t-2} + \cdots + \lambda^{m-1}\beta_1 X_{t-(m-1)} + u_t \qquad (13)$$

While (12) has a total of eleven unknown parameters, equation (13) contains only three. Of course, there is a cost: the imposed restriction may be incorrect. However, in principle this restriction can be tested. Before we address this issue, we first discuss a simplified expression of (13) (see Appendix 9.1 for a derivation). This expression is

$$Y_t = \alpha(1 - \lambda) + \beta_1 X_t + \lambda Y_{t-1} + u_t - \lambda u_{t-1} \qquad (14)$$

Note that equation (14) has only three unknown parameters. In addition, it has the advantage that only one observation is lost to accommodate the lagged effects. Further, the issue of the exact value of m is avoided. Multicollinearity should not be a problem, unless there is a strong dependency of X_t on Y_{t-1} (for example, if Y and X are sales and advertising, respectively, and advertising is strongly dependent on lagged sales).

In equation (14), β_1 is the *impact multiplier* (the instantaneous or current effect). The total effect is

$$\beta_1 + \lambda\beta_1 + \lambda^2\beta_1 + \cdots + \lambda^{m-1}\beta_1$$

This is often referred to as the *equilibrium multiplier* and it may be computed as $\beta_1/(1 - \lambda)$ (as $m \to \infty$).

In a number of empirical studies, researchers have focused on the number of periods that are needed to obtain most of the sum of current and lagged effects. For example, within a current period the proportion of the total effect realized is

$$\frac{\beta_1}{\beta_1/(1 - \lambda)} = 1 - \lambda$$

Or equivalently, the percentage of the total effect not realized within the current period is λ. Including one lagged period, these percentages are $(1 - \lambda)(1 + \lambda)$ realized, and λ^2 not realized. The table below shows these percentages in terms of λ, as well as for a specific value of λ.

Period	Percent realized	Percent not realized	Percent realized if $\lambda = 0.4$
Current	$(1 - \lambda)$	λ	0.60
One period lagged	$(1 - \lambda)(1 + \lambda)$	λ^2	0.84
Two periods lagged	$(1 - \lambda)(1 + \lambda + \lambda^2)$	λ^3	0.94

Testing the Koyck Specification

To determine whether it is appropriate to assume geometrically decaying effects for the lagged terms of the predictor variable, we can attempt to compare equation (14) with a specification that is based on the same information but does not have the restriction of geometrically declining effects. We propose one procedure that incorporates an approximation of equation (13).

Assume that we can choose an appropriate value for m. Then the sample size available for estimation of the parameters in (13) equals $n - (m - 1)$. The same sample size is then used for estimation of (14) to test the restriction of geometric decay. Ignoring the autocorrelated error terms in (14), we have

$$H_0: \beta_m = \lambda\beta_{m-1} = \lambda^2\beta_{m-2} = \ldots = \lambda^{m-1}\beta_1$$

H_A: the geometric decay restriction is false

The test-statistic that can be used is the F-test for determining the statistical significance of incremental R^2, that is,

$$F = \frac{(R^2(\mathrm{I}) - R^2(\mathrm{II}))/(\mathrm{DF(II)} - \mathrm{DF(I)})}{(1 - R^2(\mathrm{I}))/\mathrm{DF(I)}}$$

where $R^2(\mathrm{I})$ is the unadjusted R^2 value for equation (12)
$R^2(\mathrm{II})$ is the unadjusted R^2 value for equation (14)
$\mathrm{DF(I)}$ is the number of degrees of freedom left for equation (12)
$\mathrm{DF(II)}$ is the number of degrees of freedom left for equation (14)

If the null hypothesis is rejected, we conclude that the Koyck specification is false. Otherwise, we claim a lack of evidence against the Koyck specification. In that case, the full sample size $(n - 1)$ available for estimation should be used for reporting results.

The final point to be made is that the presence of an autocorrelated disturbance term in (14) makes it impossible to claim good properties for the parameter estimates obtained by applying least squares to (14). The least-squares estimates will be biased, with the nature and the extent of bias dependent upon the autocorrelation. We recommend that other procedures are considered for the estimation of parameters in a Koyck specification[2].

Partial Adjustment

Often the criterion variable captures the collective behavior of individuals. Suppose that our focus is on the sales of a particular product. The product's sales may depend strongly on its price, the price of substitutes, the availability of the product, and consumers' awareness of its existence. Now we may be able to argue that there is some type of optimal response by consumers to the specific conditions. For example, given a price decrease for the product, with other relevant conditions being held constant, at least some consumers ought to switch from a competing brand. But the actual amount of switching may be less than the optimal amount, due to a lack of awareness of the price change, or because of a lack of immediate need for the product. Thus, the observed sales may not represent the potential sales, given the price change.

To accommodate this incomplete or partial adjustment to changes in the conditions (predictor variables), we start by specifying a simple, optimal model:

$$Y_t^* = \alpha + \beta X_t \tag{15}$$

where Y_t^* is the optimal or desired behavior, and

X_t is the relevant condition that determines optimal behavior.

Due to various problems, which together we label *inertia*, the actual change in the behavior between two periods differs from the optimal change. We consider the actual change to be a fraction of the desired change, that is

$$(Y_t - Y_{t-1}) = \delta(Y_t^* - Y_{t-1}) + u_t \tag{16}$$

where δ represents a fraction, i.e. $0 < \delta < 1$.

An error term is added to the partial adjustment equation, to reflect the idea that individuals may not be consistent in their deviation from optimal behavior from one period to the next.

By substituting (15) into (16), we obtain

$$(Y_t - Y_{t-1}) = \delta(\alpha + \beta X_t - Y_{t-1}) + u_t \quad \text{or}$$

$$Y_t = \alpha\delta + \beta\delta X_t + (1 - \delta)Y_{t-1} + u_t \tag{17}$$

Equation (17) can be estimated directly from available data. We note that the observed *efect for* X_t will reflect the product of β and δ, or the product of the price effect in case of perfect adjustment and the partial adjustment effect. Thus, the current effect for X_t as estimated understates the true effect of that predictor on *optimal* behavior. Nevertheless, (17) looks a great deal like the Koyck specification.

A comparison of the Koyck specification in (14) with partial adjustment in (17) shows that different theoretical arguments can lead to virtually equal or equivalent model specifications. Of course, one remaining difference is that the error term in (17) is not expected to be autocorrelated. Therefore, parameter estimation of (17) is more straightforward, compared with the Koyck specification. In addition, if we use partial adjustment to justify the presence of a lagged criterion variable, the residuals should *not* be autocorrelated. If they are, the validity of the partial adjustment equation is in doubt.

Of course, if the adjustment is in fact complete, the parameter associated with Y_{t-1} is zero, and we can use the t-statistic for this variable to test the existence of partial adjustment.

Adaptive Expectations

The partial adjustment model is a fairly simple representation of possible behavior by individuals. It assumes that we have perfect knowledge about the relevant conditions, but imperfect behavior on the part of, say, consumers. Our knowledge may also be subject to uncertainty. Suppose that individuals make decisions about expected conditions or expected needs, that is, let

$$Y_t = \alpha + \beta X_t^* + u_t \tag{18}$$

where Y_t is the actual behavior, and

X_t^* is the expected condition or need.

Of course, the actual condition or need differs from expectations. We assume that the change in expectations is a fraction of the difference in actual and expected past values:

$$X_t^* - X_{t-1}^* = \theta(X_{t-1} - X_{t-1}^*) \tag{19}$$

where θ (theta) represents a fraction, i.e, $0 < \theta < 1$.

Then, by substituting (19) into (18) and continuing this process (see Appendix 9.2) we obtain

$$Y_t = \alpha\theta + \beta\theta X_{t-1} + (1 - \theta)Y_{t-1} + u_t - (1 - \theta)u_{t-1} \tag{20}$$

We have, again, an equation that includes a lagged criterion variable.

As with the Koyck specification, we have an error term with an autocorrelation pattern. The main difference between the Koyck model and adaptive expectations is that the predictor variable X is represented by a lagged term in the final equation in the adaptive expectations model. Essentially, all arguments and limitations relevant to the final Koyck specification also apply to the final equation for the adaptive expectations model. Specifically, when the error terms are autocorrelated, the least-squares estimates are biased. Also, the Durbin-Watson test for autocorrelation (see Chapter 7) is biased toward acceptance of the null hypothesis of independence.

As we indicated, the least-squares estimates are acceptable when the partial adjustment model is the correct justification for including a lagged criterion variable on the right-hand side of the equation. In that case, the error terms can be assumed to be independent. Yet the presence of the lagged criterion variable causes the parameter estimates to be biased. However, this bias decreases as the sample size increases.

Using a Lagged Criterion Variable to Remove Autocorrelation

In practice, researchers often add a lagged criterion variable after discovering that the residuals are autocorrelated. In this scenario, the lagged criterion variable is used not for explicit reasons (such as those specified above) but to "explain" the remaining systematic variation.

Intuitively, we can understand the explanatory power of the lagged criterion variable as follows. Suppose that Y_t depends on several predictor variables, and that at least one predictor has a trend. The result is that the criterion variable will also have a trend in its pattern of observable values. Now, if the predictor variable with the trend is omitted from the equation (a relevant variable is omitted), the residuals will still contain this trend. The result is that the residuals are autocorrelated.

Conceivably, in some situations the lagged criterion variable may serve as a proxy variable for the omitted predictor variable(s). The implicit belief then is that the inclusion of a lagged criterion variable will reduce the severity of the bias in parameter estimates. However, there is no guarantee that the parameter estimates for the other predictor variables will in fact be less biased as a result. To see this, consider the possibility that the real reason for the autocorrelation is a misspecification of the functional form. The addition of the lagged criterion variable to the model may still remove or at least reduce the amount of autocorrelation, yet the estimates are biased unless the correct functional form is used.

Aggregation

In virtually all empirical studies, the available data are aggregated over units (for example, individuals) and/or time periods. Any aggregation is likely to obscure differences in the effects. For example, if our interest is in understanding the effect of a product's price on demand, we learn the most if we obtain information about price effects separately for each individual consumer. However, the available data often consist of total sales for a market, so that we have no ability to differentiate between individuals in terms of price effects. At best, we may learn the "average" price effect for the market. In this section, we consider issues of aggregation across units as well as over the time periods. We show the nature of the expected correspondence between parameter values for separate (disaggregate) units and the parameter values that apply to the aggregate data.

We assume a simple linear model for each of two individuals:

$$Y_{1,t} = \alpha_1 + \beta_1 X_{1,t} + u_{1,t}$$

$$Y_{2,t} = \alpha_2 + \beta_2 X_{2,t} + u_{2,t}$$

The two equations do not differ in the definition of criterion and predictor variables.

However, we allow for heterogeneity in the parameters. That is, if X represents the price of a product and Y the purchase amount, the two individuals are allowed to differ in their responsiveness to price changes. In practice, we may have access only to aggregate data. For example, suppose we know the average list price faced by the two individuals for the product (the average is the same as the separate prices, if the two individuals face identical prices):

$$\bar{X}_t = \frac{\sum_{i=1}^{2} X_{i,t}}{2}$$

and we know total sales

$$Y_t = \sum_{i=1}^{2} Y_{i,t}$$

The model then may be

$$Y_t = \alpha + \beta \bar{X}_t + u_t$$

The question is, how do the parameters α and β relate to the individual parameters, α_1, α_2, β_1, and β_2? To gain insight into this, we add the two individual equations:

$$\sum_{i=1}^{2} Y_{i,t} = \sum_{i=1}^{2} \alpha_i + \sum_{i=1}^{2} \beta_i X_{i,t} + \sum_{i=1}^{2} u_{i,t}$$

But, since we use average price in our aggregate model, we modify the aggregate equation as follows:

$$Y_t = \alpha + \frac{\sum_{i=1}^{2} \beta_i X_{i,t}}{\sum_{i=1}^{2} X_{i,t}} 2\bar{X}_t + u_t$$

since $2\bar{X}_t = \sum_{i=1}^{2} X_{i,t}$

We can see from the aggregate equation that the intercept is the sum of the individual intercepts. Also, the error term is the sum of the error terms in the individual equations. As long as these error terms are independent for the two individuals, there is no reason to doubt the validity of the usual assumptions about the error term in the aggregate equation. However, if there are dependencies, the aggregate equation must incorporate whatever characteristics are responsible for these dependencies. Finally, the slope parameter in the aggregate equation is the most difficult to interpret. It is a complex expression involving the individual slope parameters, the specific values of the predictor variable faced by the individuals, and the number of individuals. Note that $\bar{X}_t$ is the predictor variable.

From this expression it is easy to see that the use of aggregate data may pose additional problems. The aggregate slope parameter is a weighted sum of the individual slope parameters, with the weights determined by the values of the predictor variable. If these weights change from one period to the next, the aggregate parameter may change as well. In addition, the number of individuals in the market may change over time. These two factors suggest that it may be inappropriate to assume constant parameters for an aggregate equation. Only if the individual parameters are equal does the aggregate equation have the properties we usually assume. This suggests that if aggregation is unavoidable, the issue of heterogeneity in parameter values should be considered carefully.

Of course, our exposition so far assumes the existence of linear effects for only one predictor variable. The correspondence between the parameters in an aggregate equation and the individual parameters becomes much more complex if we relax these two assumptions. Consider a model with two predictor variables, but linear effects, for two individuals:

$$Y_{1,t} = \alpha_1 + \beta_1 X_{1,t} + \gamma_1 Z_{1,t} + u_{1,t}$$

$$Y_{2,t} = \alpha_2 + \beta_2 X_{2,t} + \gamma_2 Z_{2,t} + u_{2,t}$$

where $Z_{i,t}$ is the number of advertising messages individual i gets exposed to in period t.

At the aggregate level, we assume the following model:

$$Y_t = \alpha + \beta\bar{X}_t + \gamma\bar{Z}_t + u_t$$

where

$$\bar{Z} = \frac{\sum_{i=1}^{2} Z_{i,t}}{2}.$$

The relationship between the parameters in the aggregate model and the parameters in the individual models is now also influenced by the correlation between $\bar{X}_t$ and $\bar{Z}_t$. If these variables are uncorrelated, we can make the same conclusion as before. (I.e., β is a weighted average of β_1 and β_2, with the weights changing if the levels of X_1 and X_2 change or the number of individuals in the market changes.) However, if $\bar{X}_t$ and $\bar{Z}_t$ are correlated, and $\gamma_1 \neq \gamma_2$, β is a function of β_1, β_2, γ_1, and γ_2. Thus, our ability to understand relationships based on aggregate data may be seriously hampered. The difficulty in understanding depends greatly on the amount of heterogeneity in the parameters at the individual level.

Our inability to gain proper understanding is also influenced by any nonlinearities that may exist. For example, suppose that the prices faced by different individuals vary (i.e., $X_{1,t} \neq X_{2,t}$). In one period, we may have $X_{1,t} = 60$ and $X_{2,t} = 70$; in another period the values may be $X_{1,t} = 55$ and $X_{2,t} = 75$. In both cases $\bar{X}_t = 65$. But the nature of the response observed at the aggregate level also depends on the variability in the prices faced by the individuals, when the relationships are nonlinear. The difficulty in gaining proper understanding depends here on the strength of the nonlinearity in the relationship at the individual level, and on the variability in the levels of the predictor variable faced by the individuals.

Aggregation over Time

It is easy to see that similar difficulties result if the observations are aggregated over time periods. For example, suppose that purchase decisions are made once a week, but data are available only on a monthly basis. If the predictor variables change in value during the month, the particular value chosen to represent the predictor variable in that month is an incomplete representation of the values faced by individuals. This problem is very similar to the difficulties considered in aggregation across individuals. We now address yet another complicating factor. Specifically, suppose that the model contains a lagged criterion variable on the right-hand side. The problems which result turn out to be independent of the specific reason for the lagged criterion variable (for example, partial adjustment or adaptive expectations). However, for ease of exposition, we assume the partial adjustment model, given its simple error-term structure.

Aggregation over Time with Lagged Criterion Variables

Suppose that we have a model with a lagged criterion variable and that we believe that the model applies to monthly data:

$$Y_t = \alpha + \beta_1 X_t + \beta_2 Y_{t-1} + u_t$$

where t refers to monthly time periods.

Assuming that this is the correct specification, we would like to use monthly data on these variables to estimate the parameters. However, if the data are available only on an annual basis, what specification should we use? One possibility is to use the identical formulation, except that the time periods are years, that is,

$$Y_T = \alpha' + \beta_1' X_T + \beta_2' Y_{T-1} + u_T$$

where T refers to annual time periods.

We should be interested in the relations between α and α', β_1 and β_1', and β_2 and β_2'.

To investigate these relations, we aggregate the specification for monthly data over twelve months.

$$\sum_{t=1}^{12} Y_t = \sum_{t=1}^{12} \alpha + \beta_1\left(\sum_{t=1}^{12} X_t\right) + \beta_2\left(\sum_{t=1}^{12} Y_{t-1}\right) + \sum_{t=1}^{12} u_t \text{ or}$$

$$Y_T = 12\alpha + \beta_1 X_T + \beta_2\left(\sum_{t=1}^{12} Y_{t-1}\right) + u_T$$

This specification is very similar to the one we specified initially for annual data. The only difference consists of the term

$$\sum_{t=1}^{12} Y_{t-1}$$

instead of Y_{T-1}. These terms are not the same. For example, if the annual data represent calendar years, Y_{1980}, or Y_T, represents the data over the months January through December in 1980. Similarly, Y_{1979}, or Y_{T-1}, represents the data over the months January through December in 1979. However,

$$\sum_{t=1}^{12} Y_{t-1}$$

is a lag of only one *month;* that is, it covers the months December 1979 through November 1980 when Y_T is Y_{1980}.

Of course, with only annual data available, it is impossible to get the data for this variable. We conclude that if the model should be specified for monthly data, the annualized version with Y_{T-1} is misspecified, causing systematic problems with all parameter estimates. Essentially, one relevant predictor variable is omitted while an irrelevant variable is included. Only if these omitted and irrelevant variables are perfectly correlated will the parameter estimates for the other variables not be systematically affected. In addition, the interpretation of the estimate of β_2' would be in error. We illustrate these concerns with two examples.

Example

In a study of commercial banks, Bryan estimated the speed with which banks adjust their reserves to changes in variables such as bank credit and money supply.[3] He estimated the parameter for lagged stock of excess reserves using several procedures, including:

1. separate estimation for each of nineteen banks using weekly data (up to 416 observations per bank);
2. aggregate estimation using weekly data, based on the sum of the observations across the banks (up to 416 aggregate observations);

3. aggregate estimation using monthly data, based on the sum of the observations across the banks (up to 96 aggregate observations).

Based on the coefficient for the lagged criterion variable he found that the results suggested speed of adjustment in reserves by these commercial banks as follows:

Estimation procedure	Coefficient	95-percent[4] closure (weeks)
1. Mean of the separate estimates	.605	3.2
2. Aggregate (weekly) estimate	.438	5.2
3. Aggregate (monthly) estimate	.099	~124.4[5]

Thus, if these commercial banks make their decisions about reserves on a weekly basis, the aggregate monthly data would be very misleading. In the example provided by Bryan, analysis of the weekly data suggests 95-percent closure of somewhere between 3 and 6 weeks. On the other hand, when monthly data are used, the analysis suggests 95-percent closure after 124 weeks! Thus, there is a substantial discrepancy in the estimated length of time required for the banks to adjust, between the weekly and monthly data.

Example

In a survey of the literature on lagged advertising effects, based on a Koyck specification (see equation (14)), Clarke computed the average value of the estimated parameters of the lagged criterion variable as follows[6]:

Data interval	Average estimated parameter	90-percent closure (months)
Weekly	0.537	0.9 months
Monthly	0.440	3.0 months
Bimonthly	0.493	9.0 months
Quarterly	0.599	25.1 months
Yearly	0.560	56.5 months

From his computations, it appears that any conclusion about the length of lagged advertising effects depends systematically on the data interval used. If weekly data are used, the average result suggests a relatively short duration for lagged effects. However, with annual data the lagged effects are estimated to continue for several years.

Recommendation

Based on the theoretical insights about aggregation effects and the empirical results when a model contains a lagged criterion variable, we recommend the following. If at all possible, disaggregate data should be used in order to have the best chance to obtain valid parameter estimates. With aggregate data, possible heterogeneity issues (e.g., differences in parameters between individual units) need to be considered. To explore issues that result in the inclusion of a lagged criterion variable, it is critical that the data interval matches the decision-making orientation of the individuals whose behavior is under study. For example, in the study of banks (reserve adjustments), we would determine how often bank managers make adjustments. In the case of advertising effects, we need to know the purchase cycle consumers use for the product under study.

Errors in Variables

The term "errors in variables" is used to indicate that the criterion variable or one or more predictor variables may be measured imprecisely. To say that a variable is measured with error does not mean that the data are systematically wrong. Instead, we suppose a variable for which the observed value deviates unpredictably from the true value. For example, if one of our variables measures the weight of individuals, a small amount of error occurs if the weight is stated in pounds. The true weight could be described in more detail by using fractions of pounds as well. However, rounding the values to the nearest pound is considered inconsequential for many studies involving individuals' weights. And, based on the observed values (stated in pounds) we cannot claim any knowledge about the direction of the rounding for individual observations. This is an example of the type of error we have in mind when we refer to "errors in variables." In this section we describe the consequences of having errors in measurement for one or more variables.

Error in the Criterion Variable

Consider first the case of having measured the criterion variable with error, that is, let

$$Y_i = Y_i^* + d_i \tag{21}$$

where Y_i is the observed or measured value,

Y_i^* is the true value, and

d_i is the measurement error.

Assume that the true value of the criterion variable is a linear function of one predictor variable:

$$Y_i^* = \alpha + \beta X_i + u_i \tag{22}$$

Then, substituting (21) into (22), we obtain

$$Y_i = \alpha + \beta X_i + u_i + d_i \tag{23}$$

Equation (23) does not seem unusual, except for the two error terms in the model. Since both errors are unobservable, they cannot be separated, but we can make some statements about the requirements for these error terms. If X_i is a random variable, it must be independent of both u_i and d_i in order for us to be able to claim desirable properties for b, the least-squares estimator. Independence will exist if the measurement error of the criterion variable is in fact random. In that case, u_i and d_i should also be independent of each other.

The main implication then is the following. Let $v_i = u_i + d_i$, the sum of the two independent error terms. Then,

$$\text{var}(v_i) = \text{var}(u_i) + \text{var}(d_i)$$

Thus, the variance of the error term in (23) is larger than the error variance in (22). Consequently, the presence of measurement error in Y_i has the effect of making it more difficult to reject the null hypothesis of no relationship between Y_i and X_i. And, the greater the variance of the measurement error, the greater the impact on the outcome of a statistical test. We conclude that any knowledge about the variance of the error in measuring the criterion variable can be used to assess a model's predictive and descriptive power with the true values of the criterion variable.

Error in the Predictor Variable

We now turn to the case where only the predictor variable is measured with error, that is, let

$$X_i = X_i^* + f_i \tag{24}$$

where X_i is the observed or measured value,
         X_i^* is the true value, and
         f_i is the measurement error.

Again, we assume a linear relation between the criterion variable and the true values of the predictor variable:

$$Y_i = \alpha + \beta X_i^* + u_i \tag{25}$$

Then, substituting (24) into (25), we get

$$Y_i = \alpha + \beta(X_i - f_i) + u_i$$

$$Y_i = \alpha + \beta X_i + u_i - \beta f_i \tag{26}$$

Equation (26) seems fairly straightforward, except for the error term. This error term has two components: u_i and βf_i. There is no reason why we cannot assume X_i and u_i

to be independent. (Note that we cannot assume X_i to be fixed, because of the random component in equation 24.) However, it is impossible to assume that X_i and f_i are independent, given that X_i is partly determined by f_i, as shown in (24). Thus, the least-squares estimate of β in (26) cannot be unbiased. The bias is in the direction of zero. And, the greater the measurement error in the predictor variable, the greater the bias of the least-squares estimate. Econometric procedures which may be used in an attempt to overcome measurement problems are available, though we do not consider them here.[7]

In summary, we have seen that the existence of measurement error in either the criterion or the predictor variable enlarges the variance of the error term in the model. This increase tends to reduce the fit of the model, and it makes it more difficult for us to observe statistically significant effects. In addition, if the measurement error pertains to a predictor variable, the least-squares estimate is biased. The magnitude of this bias depends on the variance of the measurement error. But the bias is always in the direction of zero. Thus, with measurement error in the predictor variable there are two effects to consider: (1) the variance of the estimate will be larger than otherwise, and (2) the estimate itself tends toward zero.

SUMMARY

The assumption that the predictor variables are fixed (and not random) is untenable when there is simultaneity in the relationship of interest (e.g., X depends on Y, and Y depends on X), when one of the predictor variables contains lagged values of the criterion variable, and when there is (random) error in a predictor variable. The main results which are relevant to statistical inference, however, still apply under slightly different assumptions. Primarily, this involves the assumption that the predictor variables are independent of the error term. The results in Chapter 3, for example, would still hold.

In this chapter we have discussed the difficulties resulting from simultaneous causality. For example, a least-squares analysis of Y_i as a function of X_i when X_i is also a function of Y_i, provides biased parameter estimates. We have also shown that attempts to accommodate simultaneous causality may encounter new problems, such as an inability to obtain unique estimates of all parameters (underidentification).

For lagged criterion variables we have provided three rationales. In empirical work, it is important to have an appropriate justification for the inclusion of a lagged criterion variable. The three rationales differ in the precise model specification and in the expected behavior of the error term (i.e., the presence of first-order autocorrelation). Depending on the characteristics of the error term, ordinary least-squares analysis may not be an attractive estimation procedure.

In the section on aggregation, we suggest how heterogeneity in the unknown parameters which apply to individual units may complicate the interpretation and the estimates obtained from an analysis of aggregate data. These considerations are useful for an understanding of the advantages associated with the use of disaggregate data (e.g., data on individual households as opposed to data on aggregate household behavior).

Similar arguments apply to aggregation over time periods (e.g., having annual versus weekly data).

Aggregation over time periods takes on even greater importance in the presence of lagged criterion variables. For example, if a model should ideally or theoretically be specified for weekly time periods, then it is impossible to capture the nature of the lagged effects correctly with monthly data (at least, the lagged criterion variable should not represent the previous *month's* values). Empirically, the inconsistencies between different time periods have been illustrated in studies of commercial banks' adjustments to monetary policy and in studies of lagged advertising effects on sales. Thus, the choice of the data interval requires careful consideration, especially if a lagged criterion variable is used.

The final section in this chapter deals with errors in variables. For the criterion variable, the existence of random error is a fairly minor one. Essentially, random error in the criterion variable increases the variance in the error term of the model and decreases the maximum possible value of R^2. However, for the predictor variables the implications are more severe. Random error in a predictor variable results in the slope parameter estimate being biased in the direction of a zero value.

APPENDIX 9.1
DERIVATION OF THE KOYCK SPECIFICATION

Purpose: To show how (14) results from (13).

$$Y_t = \alpha + \beta_1 X_t + \lambda\beta_1 X_{t-1} + \lambda^2\beta_1 X_{t-1} + \cdots + \lambda^{m-1}\beta_1 X_{t-(m-1)} + u_t \qquad (1.1)$$

Lag both sides of (1.1) by one period:

$$Y_{t-1} = \alpha + \beta_1 X_{t-1} + \lambda\beta_1 X_{t-2} + \cdots + \lambda^{m-1}\beta_1 X_{t-m} + u_{t-1} \qquad (1.2)$$

Multiply (1.2) by λ:

$$\lambda Y_{t-1} = \lambda\alpha + \lambda\beta_1 X_{t-1} + \lambda^2\beta_1 X_{t-2} + \cdots + \lambda^m\beta_1 X_{t-m} + \lambda u_{t-1} \qquad (1.3)$$

Subtract (1.3) from (1.1):

$$Y_t - \lambda Y_{t-1} = \alpha - \lambda\alpha + \beta_1 X_t - \lambda^m\beta_1 X_{t-m} + u_t - \lambda u_{t-1} \qquad (1.4)$$

As $m \to \infty$, $\lambda^m \to 0 (0 \le \lambda < 1)$, and hence

$$Y_t = \alpha(1 - \lambda) + \beta_1 X_t + \lambda Y_{t-1} + u_t - \lambda u_{t-1} \qquad (1.5)$$

Based on the value of λ we can specify the effects of all lagged values of X on Y. Note, however, that the error term has a unique structure. In fact, this is one example where the error term is not expected to be independent. The estimation procedure for (1.5) has to reflect this.

APPENDIX 9.2
DERIVATION OF THE ADAPTIVE EXPECTATIONS MODEL

Purpose: To show how (20) results from substituting (19) into (18).

A restatement of (19) is

$$X_t^* = X_{t-1}^* + \theta(X_{t-1} - X_{t-1}^*)$$
$$= \theta X_{t-1} + (1 - \theta)X_{t-1}^*$$

Substituting this equation for X_t^* into (18) gives

$$Y_t = \alpha + \beta(\theta X_{t-1} + (1 - \theta)X_{t-1}^*) + u_t$$

By continuing to substitute for the expected condition or need, we obtain

$$Y_t = \alpha + \beta\theta X_{t-1} + \beta\theta(1 - \theta)X_{t-2} + \beta\theta(1 - \theta)^2 X_{t-3}$$
$$+ \cdots + \beta\theta(1 - \theta)^{m-1}X_{t-m} + u_t$$

This result is very similar to the assumption of geometrically declining weights in the Koyck specification. Using the same operations performed in Appendix 9.1, we get

$$Y_t = \alpha\theta + \beta\theta X_{t-1} + (1 - \theta)Y_{t-1} + u_t - (1 - \theta)u_{t-1}$$

APPENDIX 9.3
COMPUTATION OF 95-PERCENT CLOSURE

Suppose that, for a variety of reasons, we do not adjust optimally to changes in conditions. For example, let

$$Y_t - Y_{t-1} = \delta(Y_t^* - Y_{t-1})$$

where $0 < \delta < 1$

That is, the actual change in the value of the criterion variable is assumed to be a fraction of the desired change (Y_t^* is the optimal value of Y in period t). For example, let $Y_{t-1} = 100$, and suppose that, because of changes in other conditions, the optimal value for $Y_t^* = 120$. Then, if $\delta = 0.6$ we would observe $Y_t = 112$, that is,

$$Y_t - 100 = 0.6(120 - 100)$$

$$Y_t = 112$$

Now, let $Y_{t+1}^* = 120$, that is, assume that the optimal value does not change from t to $(t + 1)$. Then, with $\delta = 0.6$, we will next have

$$Y_{t+1} - Y_t = 0.6(Y_{t+1}^* - Y_t) \quad \text{or}$$

$$Y_{t+1} - 112 = 0.6(120 - 112)$$

$$Y_{t+1} = 116.8$$

Thus, under equilibrium conditions, in the first period we obtain 60-percent closure (with $\delta = 0.6$), and in the second period we increase the closure to 84 percent, i.e.:

$$\frac{Y_{t+1} - Y_{t-1}}{Y_{t+1}^* - Y_{t-1}} = \frac{16.8}{20}$$

The question addressed by Bryant is, how many periods does it take to get 95-percent closure?

To examine this, note that, after two periods we have $\delta + \delta(1 - \delta)$ closure, or $0.6 + (0.6)(0.4)$ or 84-percent closure. After three periods, we have $(\delta + \delta(1 - \delta) + \delta(1 - \delta)^2)$ closure, or $0.6 + (0.6)(0.4) + (0.6)(0.16))$, or 93.66-percent closure. Conversely, after one period we are left with $(1 - \delta)$ nonclosure, after two periods $(1 - \delta)^2$, and after three periods $(1 - \delta)^3$ nonclosure. To find 95-percent closure, we want 5 percent nonclosure, or

$$(1 - \delta)^q = 0.05 \quad \text{or}$$

$$q \ln(1 - \delta) = \ln 0.05$$

For $\delta = 0.6$, we have

$$q \ln 0.40 = \ln 0.05$$

$$q = \frac{\ln 0.05}{\ln 0.40} = \frac{-2.99573}{-0.91629} = 3.27$$

Or with a coefficient for the lagged criterion variable of 0.6, 95-percent closure is achieved after 3.27 periods.

In general, it is desirable to have a large amount of closure quickly. Often, we consider how many periods it takes to achieve 90 or 95-percent closure. In the example discussed above, a little more than 3 periods are required to obtain close to optimal adjustment to changes in conditions.

ENDNOTES

1. L. M. Koyck, *Distributed Lags and Investment Analysis* (Amsterdam: North-Holland, 1954).

3. See, for example, J. Johnston, *Econometric Methods,* 3rd ed. (New York: McGraw: Hill, 1984).

3. Bryan, William R., "Bank Adjustments to Monetary Policy: Alternative Estimates of the Lag." *American Economic Review*, 57 (September 1967): 855–864.

4. See Appendix 9.3 for an explanation.

5. Bryan reports this result as 28.7 months.

6. Clarke, Darral G., "Econometric Measurement of the Duration of Advertising Effect on Sales," *Journal of Marketing Research*, 13 (November 1976): 345–57.

7. See, for example, G. C. Chow, *Demand for Automobiles in the United States: A Study in Consumer Durables* (Amsterdam: North-Holland, 1957); and Y. Haitotsky, "On Errors of Measurement in Regression Analysis in Economics," *Review of International Statistical Institute*, (1972): 23–35.

REFERENCES

Bryan, William R. "Bank Adjustments to Monetary Policy: Alternative Estimates of the Lag." *American Economic Review*, 57 (September 1967): 855–864.

Chow, G. L. *Demand for Automobiles in the United States: A Study in Consumer Durables*. Amsterdam: North-Holland, 1957.

Clarke, Darral G. "Econometric Measurement of the Duration of Advertising Effect on Sales." *Journal of Marketing Research,* 13 (November 1976): 345–57.

Haitovsky, Y. "On Errors of Measurement in Regression Analysis in Economics." *Review of International Statistical Institute*, (1972): 23–35.

Johnston, J. *Econometric Methods*. 3rd ed. New York: McGraw-Hill, 1984.

Koyck, L. M. *Distributed Lags and Investment Analysis*. Amsterdam: North-Holland, 1954.

10

The Conduct and Evaluation of Regression Applications

In this chapter, we discuss critical aspects in the conduct and evaluation of regression analysis. Our objective is to provide guidelines that can be used to carry out projects as well as to determine the quality of the work done by others. After putting the content of this chapter in perspective, we provide a general procedure for the application of regression analysis. Within this general procedure we make reference to earlier chapters, and we introduce some additional topics.

Before we can discuss the elements of a procedure for the application of regression analysis, we need to define the main purpose of the study. Although regression analysis is also used for predictions or forecasts, we have emphasized the objective of description or understanding. Of course, if the final model provides a valid description of the effects which the predictor variables have on a criterion variable, the model should be useful for predictions as well. Nevertheless, it is often possible to generate models that have predictive accuracy superior to that of the best descriptive model. Models developed purely for forecasting, however, would not provide the insight we aim to show.

If forecasting is the primary purpose, the approach is often to use the most basic description of systematic variation in the criterion variable. However, the effort usually does not consider the reason(s) for the systematic variation. On the other hand, for valid description it is imperative that the model provides satisfactory explanation of the systematic variation.

Although we make the distinction between prediction and understanding, in many studies both purposes play a role. And ultimately, the goal is often to use the regression results for *prescriptions*. That is, based on the estimated effects for the predictor variables in the model, we want to be able to choose levels of one or more predictor variables that will produce desired results for the criterion variable. Of course, if the model is an incorrect description of the relations postulated, we will not be able to choose appropriate values for the predictors correctly. In practice, we can never be sure that our model is correct, but there are many procedures and tests available to us for determining whether a given model has shortcomings. To do this, we have to rely heavily on statistical tests and other procedures that allow us to discriminate between alternative model specifications as well as to see if a given model should be rejected altogether.

Users of regression analysis have only recently begun to recognize the importance of inspecting the residuals. Simply stated, we should question a model's validity if the residuals exhibit systematic variation. In the field of forecasting, most available procedures have the objective of exploiting all systematic patterns in the criterion variable, so that the final model is by definition acceptable in the sense that it only leaves "white noise" or random variation unaccounted for. However, when our purpose is to describe or understand, we do not have the liberty to provide arbitrary, mechanical explanations of systematic variation in the criterion variable. Instead, we are constrained by the ability of the predictor variables to account for all systematic variation. Thus, the more complete our theoretical understanding of the problem we are studying, the better our chances of identifying all relevant predictor variables.

Historically, empirical work based on regression analysis has focused heavily on the magnitude of summary statistics. In particular, unadjusted or adjusted R^2 values have often served as the principal measure of a model's quality. For example, some users are known to arbitrarily reject any result based on a model with an R^2 value less than 80 percent. We have pointed out some of the difficulties associated with R^2 values in Chapter 8.

Apart from the heavy emphasis on R^2 values, users have taken a very narrow interpretation of the most commonly used test statistic involving the residuals. When the observations represent a single time series, the null hypothesis of independence for the error term is usually tested with the Durbin-Watson test. If the null hypothesis is rejected, researchers tend to interpret this as indicating only that the estimated standard errors are biased. This interpretation appears to be derived from the typical exposition of first-order autocorrelation in standard econometrics texts. The exposition goes along the following lines. Assume that the error terms have zero expected value but are not independent. Then it can be shown that the least-squares estimates are unbiased. But, given that the estimated standard errors are not unbiased, it is better to make a technical correction for the autocorrelation in order to get asymptotically unbiased standard errors.

Now, one difficulty with this treatment of autocorrelated error terms is that there is no indication provided for the source of the autocorrelation. Thus, although it is possible for the model to be correctly specified and for the error term to be autocorrelated, in practice it is more likely that autocorrelation results from model misspecification. The authors of standard econometrics texts have not said that the model can be assumed to be correct when autocorrelation is observed, but the textbook treatment may have contributed to this common conclusion. We argue that autocorrelation should be interpreted as an indication of model misspecification (e.g., omitted variable, wrong functional form), *unless* a plausible reason for the lack of independence in the error term is specified prior to data analysis. For example, it is easy to show that an erroneous functional form (see Chapter 7) or an omitted predictor variable can lead to autocorrelation in the residuals. Given a finding of autocorrelation in the residuals, it is especially important to investigate whether there is any evidence of omitted variables or functional form problems.

Much empirical work has relied upon stepwise regression analysis. Someone has characterized the user of stepwise regression as a person who checks his or her brain at the entrance of the computer center. Indeed, although stepwise regression analysis may have some useful characteristics, it has many undesirable features. The use of stepwise regression procedures requires only the specification of a criterion variable and a number of predictor variables. A forward selection procedure for stepwise regression starts with a simple regression analysis of the criterion variable and the predictor variable with the highest simple correlation coefficient. In the second stage, a multiple regression analysis is done, using the predictor variable included in the first stage and the predictor variable that has the highest added explanatory power (given that the first predictor is already in the model). This procedure is continued until some criterion is optimized. For example, the best model may be determined based on the maximization of adjusted R^2. It is also possible to do a backward stepwise regression analysis by first including all predictor variables and deleting one predictor variable at a time, based on, say, the absolute magnitude of the t-ratios. However, even if the same criteria are used for forward and backward stepwise regression analyses, there is no guarantee that the solutions are identical.

Perhaps the most serious limitation associated with stepwise regression analysis is that all statistical tests are inapplicable, strictly speaking. The statistical procedures require a priori specification of a model; in stepwise regression analysis, the data are used to determine the model. In addition, it is unusual for users of stepwise regression analysis to consider the appropriateness of the functional form of the relations, and diagnostic tests based on residuals are rarely carried out. Thus, the result of a stepwise regression analysis is likely to overstate the model's explanatory power, and the user may fail to recognize the possible existence of systematic variation in the residuals.

Yet there are some positive aspects associated with stepwise regression analysis. The procedure is often used when there are many potentially relevant predictor variables. And, given that omitting a relevant variable is likely to bias the estimated effects, it is better to err on the side of having too many than having too few predictors. As we have seen, a superfluous variable is simply expected to obtain an insignificant effect. In addition, recent simulations have provided new tables that can be used to determine the statistical significance of stepwise regression results (e.g., McIntyre et al.[1]).

Unfortunately, however, there is a tendency for regression users to hide much of the data analysis. For example, even if stepwise regression is used, or if hundreds of regressions are run, the user may report only the final result and pretend that only one regression analysis was conducted. Obviously, hiding the actual procedures used makes it likely for the user to overstate results, causing all other users to be disappointed when the results are applied to making decisions. We recommend, therefore, that all reports include precise and complete information about the procedures used. However, we also recommend that users specify appropriate functional forms and use diagnostics to verify the assumptions. Algorithms such as stepwise regression analysis should be reserved for situations where the research is entirely exploratory, and where the researcher has extreme difficulty justifying any model specification prior to data analysis.

Below, we discuss desirable elements of a procedure for the systematic specification, estimation, and testing of a model for regression analysis. Our procedure assumes that the researcher has relevant knowledge about the problem or that such knowledge can be obtained from others.

Use Relevant Theory to Specify the Initial Model

The first step in our recommended procedure is to determine all relevant knowledge to the problem and to use this knowledge to specify an initial model. This knowledge should be useful both for the specification of potentially relevant predictor variables and for the specification of plausible functional forms. Generally, it is better to err on the side of specifying too many rather than too few predictor variables. As we have argued, irrelevant predictors have a zero expected value for the slope. On the other hand, an omitted variable is likely to bias the estimated effects for the included predictor variables. Specifying the set of predictor variables does not imply that each one is necessarily relevant. Rather, we have some reason for expecting an effect, and the results may either support or disprove our expectations. Specifically, the relevance of each individual predictor variable in a multiple regression analysis can be determined using the t-test (assuming, of course, that the assumptions required for the test are met).

It is often easier to specify potentially relevant predictor variables than it is to specify appropriate functional forms for each predictor variable. In general, the more uncertainty about the functional form, the more reason there is to compare alternatives and to use the data to select one alternative. For example, uncertainty about linear versus quadratic functional forms should result in the addition of the squared values of the predictor as a separate variable. The t-ratio for the coefficient of the squared variable can be used to choose between the linear and quadratic formulations. Ideally, such alternatives are specified in advance, based on knowledge about the problem. If only one functional form is considered, and if the residuals suggest that the functional form is inadequate, the researcher has to reconsider the theory and relevant prior knowledge. The data usually are not precise enough to indicate the appropriate functional form.

Consider Choices in the Measurement of the Criterion Variable

It may be that there is little flexibility in the definition or measurement of the criterion variable. But it is important for users of regression analysis to consider alternatives. For example, if the criterion variable involves dollar values over time (e.g., profits, revenues), there is the choice of using current or constant dollar values. If one or more predictor variables are also measured in dollars, the use of current dollars may result in artificially high explanatory power due to the common element of inflation on both sides of the equation. In addition, if more than one predictor variable is measured in current dollars, the common element of inflation may increase the degree of multicollinearity.

Similarly, if variables are defined over a population of individuals, a model may show artificially high explanatory power, if the population is growing over time and time-series data are used. Thus, it is preferable to define such variables on a per capita basis, both to eliminate the effect of population growth and to minimize multicollinearity.

For the measurement of brand performance, many regression users employ market share (in units or in revenue). An alternative measure consists of sales volume. Here the choice is between a relative (market share) and an absolute (sales volume) criterion variable. However, apart from that, there may be implicit assumptions that make one measurement alternative more attractive than another. For example, with sales volume the researcher has the opportunity to specify predictor variables, such as prices, for many possible competitors. On the other hand, the use of market share involves the implicit assumption that the relevant set of competitors can be correctly specified independent of the data. After all, to measure market share we need to be able to define the market. In addition, the use of market share can be shown to involve other implicit assumptions. Thus, even if management has decided that market share is a critical criterion variable, the regression analysis may still benefit from a focus on sales volume. Market share can still be obtained indirectly by using independent estimates of product category sales. In this manner, the validity of the regression results is not affected by using market share as the criterion variable.

We have also emphasized the sensitivity of the results to the choice of time periods. In the case of time-series data, results may differ systematically between, say, weekly and monthly data. In general, it appears to be preferable to use data at as disaggregate a level as possible.

Measurement of Predictor Variables

Similar arguments apply to the measurement of the predictor variables. For example, if we are interested in learning the effect of advertising on a firm's profits or sales of a product, we may measure the advertising variable as expenditures or number of exposures. Neither of these measures captures fully the multidimensionality of advertising effects. Nevertheless, one measure may have more desirable properties than another depending upon the study objective.

Sources of Variation

We have stressed the influence which the amount of sample variation in a predictor variable has on the reliability of parameter estimates. In addition, greater variation provides better opportunities to test the appropriateness of the functional form.

If we consider sample variation in the criterion variable, it is useful to note that both the amount and reason for variation may depend on the nature of the data. For example, cross-sectional variation often differs systematically from time-series variation. It is likely, therefore, that the model specification (e.g., the set of relevant predictor variables) differs between time-series and cross-sectional data.

Obtain Agreement about the Initial Specification

It is important that clients for whom a regression analysis is done agree prior to data analysis and results about the model specification. The primary reason for this is that some results may contradict intuitive expectations. Some people find it very difficult or impossible to accept such results. They will then argue that the researcher must have missed some relevant variable or made some other mistake. If the client agrees to the a priori specification, such objections are less likely to occur.

A second reason for checking with clients is that they are likely to have knowledge relevant to the problem. This knowledge should be used in specifying the criterion and predictor variables, and in generating alternative models to be compared and tested against each other.

Check the Data

Having done everything possible to justify the initial model prior to data analysis, it is now important to make some basic checks on the data. After placing the data in a computer file, obtain summary statistics to verify the general accuracy of the data. For example, we can examine the means and standard deviations of the criterion and predictor variables and determine whether these values are reasonable in our judgment. In addition we can use "exception reports" to determine the possibility of errors in the recording or coding of data.

Test the Initial Full Model

Suppose that we have specified one model that includes all potentially relevant predictor variables with the most plausible functional forms. Our attention should now be directed toward estimating and testing this full model. The F-test (see Chapter 5) can be used to determine whether this model's explanatory power is better than could have been obtained by chance. Essentially the test statistic allows us to discriminate between two alternatives: the null hypothesis, which states that the predictor variables together are irrelevant, and the alternative hypothesis, which states that at least one predictor variable is relevant. This test must be carried out on the initial model to avoid the bias introduced if the data are used to construct the model (as in stepwise regression).

Test Alternative Models

If the null hypothesis that all predictor variables are irrelevant is rejected, it is possible to compare alternative model specifications. For example, if there is uncertainty about the functional form for a given predictor variable, and if the full model includes the squared values as a separate predictor variable, the t-test may be used for the null hypothesis that the squared term is irrelevant. If the null hypothesis of a zero parameter for the squared term is not rejected, this term can be deleted from the model. The consequence of this result is then that a reduced model (the full model without the squared term) is favored over the full model.

We note that such specific tests, involving possible simplifications of the model, are often carried out one at a time. For example, if there are squared terms for two different predictor variables, multicollinearity may prevent either one from being statistically significant, even if together they increase the model's explanatory power by a significant amount. An incremental F-test could be used to test the null hypothesis that both terms together are irrelevant. If this is not done, and if both squared terms have low t-ratios, one term should be deleted and another multiple regression analysis run to see if the t-ratio for the remaining squared term changes dramatically. (See the discussion about collinearity between predictor variables in Chapter 4.)

Tests of the relevance of one or more predictor variables are the easiest to make. Such tests are called *nested* tests, because one model is a simplification of another. Often, however, we want to compare alternative models which may have the same number of predictor variables. For example, a given predictor may be assumed to have linear effects in one model and nonlinear effects (perhaps using a logarithmic transformation) in another. Such comparisons may be made by constructing a supermodel (see Chapter 7) that includes both the linear and nonlinear effects. Each alternative model specification can be tested against this supermodel. Ideally, the test results are such that one alternative is as good as the supermodel, while the other one is inferior. In that case, the result is unambiguously in favor of the chosen alternative. But it is also possible for the supermodel to be significantly better than either alternative. This outcome suggests that neither functional form is adequate by itself, and that additional work is required to identify a plausible functional form. Finally, if the supermodel is not significantly better than either alternative, the data are not capable of providing a statistically significant distinction between the two alternatives.

It is perhaps clear from this discussion that nested models may be preferred, because with them we can reject one model in favor of another. No matter what the outcome of a test is, then, we have one alternative model available. The use of a supermodel, however, could leave us with the result that neither of the alternative models is acceptable. Although such a result may be uncomfortable (e.g., there is no guidance provided to identify other model specifications), it is important to use all statistical procedures available for the most detailed and comprehensive scrutiny of the proposed model(s).

Show the Fragility of Parameter Estimates

Especially if there are several models that are approximately equally plausible, it is appropriate to determine whether the conclusions about key effects (parameter estimates) vary dramatically between alternative models. Leamer has suggested that applied regression analysis should include explicit examination of the sensitivity of the conclusion about relationships to the inclusion of other variables.[2] The less sensitive (fragile) a given parameter estimate is to the specification of alternative models, the more confidence we can have in the result. We should be especially interested in examining the fragility if (1) we are primarily interested in estimating the parameter for one predictor variable, (2) we are unable to discriminate unambiguously between alternative model specifications containing this predictor, and (3) our theoretical knowledge about the problem is quite limited.

Examine the Residuals

We assume that at this point there is one model specification that survived whatever statistical tests have been made so far. We should now determine whether there is any evidence against this particular model. For this we can use the diagnostic procedures discussed in Chapter 7. These procedures may also result in some discomfort, in the sense that systematic variation remaining in the residuals should cause us to question the validity of the model (e.g., if the Durbin-Watson statistic indicates a lack of independence based on a time series of residuals).

Inspect the Equation for Logical Inconsistencies

It is also useful to determine whether the final model is acceptable in terms of its implications. Consider the problem of estimating the relationship between the percentage of households with touch-tone dialing and the length of time the touch-tone service has been available (see Chapter 6). By assuming a linear relation between the two variables, we might obtain a result such as the following:

$$\hat{Y}_i = 21 + 1X_i$$

where Y_i is the percentage of households with touch-tone dialing in territory i, and

X_i is the number of months the service has been available in territory i.

This equation implies that the percentage of households with touch-tone dialing is predicted to be 21 percent when the service is not available. After having had the service available for 100 months, the percentage is predicted to be 121 percent.

Clearly, both implications are unacceptable. Regardless of the model's R^2 value or any other aspect, the equation is misleading. Now it is possible that the linear equation provides a good representation of the relationship for the range of data available. For example, suppose that the sales territories varied in touch-tone availability from 18 to 45 months. Then the predicted percentages for the sample data range from 39 to 66 percent. But unless we specify that the equation should not be used outside this interval, the result is likely to be misused. Thus, we have the options of (1) stating the result as follows,

$$\hat{Y}_i = 21 + 1X_i \qquad \text{for } 18 \leq X_i \leq 45$$

or (2) finding a more general result. We strongly prefer the second option. For example, what are we to do with territories for which the value of X exceeds 45? Note also that the specification of the range is arbitrary. If the equation works when $X = 45$, why would it not work when $X = 46$?

To inspect the validity of the equation, we recommend that predictions are obtained for a variety of extreme values of the predictor variables. For each prediction, it is important that our knowledge and understanding of the problem is consistent with these predictions.

Use a Validation Sample

After everything possible has been done to test the model against the data used for parameter estimation, it is often useful to test the model's validity on new data. Some researchers make a habit of splitting the available data into two parts. One subset is used to develop the model and to estimate parameters. The other subset (the *holdout* sample) is used to get a measure of the model's descriptive or predictive validity. We use the term *validation* sample to refer to an independent sample of new data. We describe below three reasons for using a validation sample and discuss examples of measures that can be used to estimate the model's accuracy.

1. A regression model is usually developed based on a number of iterations, or different runs. In principle, many different regression runs may be made, each one using the same data, until one model specification is chosen. As the number of iterations increases, the likelihood that the results are inflated also increases. For example, the adjusted R^2 or the standard deviation of residuals only considers the number of parameters estimated (and the number of observations) in the final equation. There is no adjustment for the number of iterations used before this final model is chosen. In an extreme situation, stepwise regression analysis may have been used to generate the model. Therefore, a validation sample should be used to get a better (e.g., a less biased) estimate of the final model's descriptive or predictive validity. If the model is used for predictions, the uncertainty associated with a particular prediction should be assessed with the estimated standard error of that prediction (see Chapter 2). The formula incorporates, among other things, the standard deviation of the residuals. If this measure understates the true variability in the error term, the estimated prediction standard errors will also be biased downward.

2. Sometimes we use a summary statistic to describe a model's predictive ability. One measure used for this is the model's adjusted R^2 value. Now, the adjusted R^2 value is a virtually unbiased estimate of the explanatory power associated with the predictor variables in the model, assuming no iterations in model development. However, the parameter estimates reflect both the true, but unknown, parameter values in the population and the "noise" present in the estimation sample. Of course, the noise present in one sample is expected to be independent of the noise present in another sample. For that reason, the model's predictive validity (as measured by adjusted R^2) for another sample is overstated. In addition, because R^2 is affected by the amount of variation present in a sample (see Chapter 8), there is another source of uncertainty about the model's validity for another sample.

3. A model is always an incomplete representation of reality. Any relevant factor outside the model (i.e., an omitted variable) which changes in impact over time or across observational units may affect the model's validity. The greater the differences in such outside factors between estimation and validation samples, the better the opportunity to test the robustness of the model. If the validity of the model is substantially smaller in the validation sample than was estimated based on the estimation sample, the extent to which we can generalize the final model is in question. It is important in this regard that a substantial decrease is compared with the differences in characteristics between estimation and validation samples. Such differences may point to nonlinearities or interaction effects not presently accommodated.

We have already hinted, especially in discussing the third reason for using a validation sample, that a validation test is most powerful when there are substantial differences in the characteristics or circumstances between estimation and validation samples. Consider now the procedures used to construct a validation sample; we review this separately for cross-sectional and time-series data.

Cross-Sectional Data

With cross-sectional data, it is common for researchers to split the total sample randomly into estimation and validation samples. For example, seventy percent of the observations could be randomly chosen to form the estimation sample, leaving the remaining thirty percent for validation. To determine the value of this procedure, we can check to see which of the three reasons for using a validation sample may be satisfied with this procedure. It appears that reasons 1 and 2 are handled by this procedure. For example, any capitalization on chance (random noise) in the estimation sample will be negated in the validation sample. This should be true even though the characteristics of the estimation and validation samples are expected to be the same (because of random sampling). On the other hand, the third reason is clearly not addressed, precisely because the characteristics of the two subsamples are expected to be identical.

Time-Series Data

Researchers using a single time series do not randomly choose observations for the estimation sample. Instead, they choose the earliest observations, leaving the most recent data for validation. This procedure can satisfy all three reasons, especially if there are substantial changes over time. For example, if an econometric model does not incorporate aspects of the economy, and the economic conditions differ substantially between the two subsamples, we have an opportunity to determine the model's robustness under varying economic conditions. Thus, the greater the differences between the two time periods, the more powerful this validation test is.

These considerations suggest that it is useful to consider other ways of splitting the data into two subsamples. In particular, we may propose to determine the model's robustness under a variety of conditions. Thus, if we are interested in testing the model under different economic conditions, we may want to split the data such that the conditions vary as much as possible. Of course, we can also incorporate these economic conditions explicitly into our model, but there are many ways in which that can be done. If there is great uncertainty about how such economic variables should be included, if at all, it may be preferable to first examine the issue in a general manner.

For cross-sectional data, we recommend that the data not be randomly divided into estimation and validation samples. Instead, we should identify characteristics of the data that are most likely to be ill-represented in the model. For example, there may be nonlinearities or interactions that have been ignored. Probably the best procedure to explore this is to split the data into two equal-sized subsamples. If the focus is on one predictor variable, the subsamples can be created by placing the observations with the lowest values on that variable in one subsample and the remaining observations in the other subsample. A comparison of the parameter estimates across the two samples should suggest whether the model needs to be expanded. This comparison can be based on a test for the statistical significance of the difference between the parameters for the predictor variable used for sorting (a test of nonlinearity) or between parameters for the other predictor variables (a test of interaction) or both. The *F*-test for pooling data (see Chapter 8) can be used for this procedure.

Validation Measures

The validity of the final model is usually assessed in terms of predictive accuracy. That is, the estimated equation (from the estimation sample) is used to predict values for the criterion variable in the validation samples. These predicted values are then compared with the actual values. Note that in this case the observations in the validation sample are *not* used to estimate parameters. Instead, the parameters estimated in the estimation sample are used to obtain the predicted values in the validation sample.

Average Prediction Error

If it is of interest to see whether the predictions are correct, on the average, we may compute the average prediction error:

$$APE = \frac{\sum_{j=1}^{n} (Y_j - \hat{Y}_j)}{n}$$

where APE is average prediction error,

Y_j is the actual value for the jth observation of the criterion variable,

$\hat{Y}_j$ is the corresponding predicted value, and

n is the validation sample size.

Note that, although the average residual in the estimation sample is guaranteed to be zero, as long as the equation contains an intercept (see Chapter 2), this is no guarantee that the predictions have the same value, on the average, in the validation sample. Note also that the denominator in the APE formula is the actual number of observations in the validation sample, because the observations are not used to estimate parameters. If desired, we can test the null hypothesis that the mean prediction error is zero, based on the t-test for the mean. Inability to reject the null hypothesis means that there is no evidence of bias in the predictions. It does not indicate anything about the accuracy of the predictions, however.

Average Absolute Prediction Error

A simple and easily computed measure of predictive accuracy is based on the average absolute difference between actual and predicted values:

$$AAPE = \frac{\sum_{j=1}^{n} \left| Y_j - \hat{Y}_j \right|}{n}$$

where AAPE is average absolute prediction error.

This measure is appropriate when the cost of forecast errors does not depend on the direction or the magnitude of the errors. However, given that the objective of regression analysis is to minimize the sum of squared deviations between actual and fitted values in the estimation sample, it would be inconsistent to evaluate the quality of the equation based on the absolute differences in the validation sample.

Average Squared Prediction Error

To overcome the objection to the second summary measure, we can measure the differences in terms of squared deviations:

$$\text{ASPE} = \frac{\sum_{j=1}^{n} (Y_j - \hat{Y}_j)^2}{n}$$

where   ASPE is average squared prediction error.

Intuitively, we can say that the use of squared terms involves an extra penalty for large prediction errors, without regard to the direction of the errors. Although this measure is consistent with the objective of regression analysis, it has the drawback that it summarizes the differences in squared units.

Root Average Squared Prediction Error

To get the summary measure stated in the units of measurement for the criterion variable, we take the square root of ASPE:

$$\text{RASPE} = \sqrt{\frac{\sum_{j=1}^{n} (Y_j - \hat{Y}_j)^2}{n}}$$

where   RASPE is root average squared prediction error.

The result for the validation sample can be compared to the standard deviation of residuals computed for the estimation sample. In general, we expect the value of RASPE to be greater than the standard deviation of the residuals. The magnitude of the difference is influenced by several factors, primarily the three we identified earlier as reasons for using a validation sample.

Correlation Coefficient (r)

If we prefer to have a relative measure of correspondence between actual and predicted values in the validation sample, we may compute the simple correlation coefficient. The squared correlation coefficient can be compared with the adjusted R^2 value obtained in the estimation sample. We expect the squared correlation coefficient to be lower than this adjusted R^2 value for the same reasons that we expect the root average squared prediction error to be higher than the standard deviation of residuals. The comparison between the squared correlation coefficient and the adjusted R^2 is also subject to the limitations discussed in Chapter 8. For example, a difference in the amount of variation in the criterion or predictor variables between the two subsamples influences the comparability. In addition, it is important to realize that the magnitude of the correlation coefficient is not affected by any bias that may exist (e.g., as measured by APE).

Theil's Decomposition

Theil has proposed a decomposition of ASPE to determine how much of the prediction error is due to bias, the regression, and the disturbance[3]:

$$ASPE = APE^2 + (s_{\hat{Y}} - rs_Y)^2 + (1 - r^2)s_{\hat{Y}}^2$$

where $s_{\hat{Y}}$ is the standard deviation of the predicted values,

s_Y is the standard deviation of the actual values,

r is the correlation coefficient between actual and predicted values, and

ASPE and APE are as defined previously.

The first part of the decomposition captures the bias. The second part is about the extent to which the relationship is systematically different in the validation sample compared with the estimation sample, while the third part captures the random noise or disturbance in the criterion variable. If the model is valid, the bias and regression parts should each be close to zero.

For time-series data it may also be useful to compare the predictive validity of the model with a naive approach, by using Theil's U-statistic:

$$U = \sqrt{\frac{\sum_{t=2}^{n} (Y_{t+1} - \hat{Y}_{t+1})^2}{\sum_{t=2}^{n} (Y_{t+1} - Y_t)^2}}$$

The denominator in the U-statistic measures the squared differences between the successive actual values for the criterion variable. The numerator measures the squared differences between the actual and predicted values. If the numerator exceeds the denominator ($U > 1$), our model is worse than a naive approach which predicts the value of Y in period $t + 1$ to be the value that occurred in period t. If our model is useful for forecasting purposes, even for just one period ahead, our model should outperform such a naive approach. Only if $U < 1$ does our model outperform this naive approach. Note that the use of squared differences is consistent with the least-squares criterion of regression analysis. We emphasize, however, that this statistic can only be used with time-series data.

Finally, it is appropriate to point out that the choice of one or more measures should depend on the purpose of the study. If it is possible to link prediction errors directly to a managerial aspect, still other summary measures can be developed. For example, suppose that a prediction error is serious only if the absolute magnitude exceeds some specified value. We can then count the number or proportion of times the prediction error in the validation sample exceeds this value. Alternatively, it may be possible to express any forecast error in terms of the cost in dollars. This may be especially appropriate if several alternative actions are being considered (for example, the size of a plant or the magnitude of an investment), and the forecast errors are expressed in costs separately for each alternative. The point here is that the appropriateness of any summary measure depends on the particular use or application of the model.

SUMMARY

In this chapter, we have proposed a series of steps that may be used for the conduct and evaluation of regression analysis. The proposed list of steps is not comprehensive in covering all aspects that may be relevant to a given project. However, we have identified what we believe to be essential considerations in empirical work based on regression analysis. In the next chapter, we discuss in detail the application of regression analysis in one particular study.

ENDNOTES

1. McIntyre, Shelby H., David B. Montgomery, V. Srinivasan, and Barton A. Weitz, "Evaluating the Statistical Significance of Models Developed by Stepwise Regression," *Journal of Marketing Research*, 20 (February 1975): 1–11.

2. Leamer, Edward, "Sensitivity Analysis Would Help," *American Economic Review*, 75 (June 1985): 308–13.

3. Theil, Henri, *Applied Economic Forecasting* (Amsterdam: North Holland, 1971).

REFERENCES

Leamer, Edward, "Sensitivity Analysis Would Help," *American Economic Review,* 75 (June 1985): 308–13.

McIntyre, Shelby H., David B. Montgomery, V. Srinivasan, and Barton A. Weitz, "Evaluating the Statistical Significance of Models Developed by Stepwise Regression," *Journal of Marketing Research,* 20 (February 1975): 1–11.

Theil, Henri, *Applied Economic Forecasting* (Amsterdam: North Holland, 1971).

11

Regression Analysis in Practice: Predicting Academic Performance of MBA Program Applicants

Now that we have discussed the essential components of empirical work based on regression analysis, we present a detailed application. This application was carried out to aid and strengthen the MBA admissions process at Stanford University.[1] In this chapter we discuss the motivation for the study, the role of the model to predict applicants' academic performance, the measurement of academic performance, the identification and measurement of predictor variables, model development, the testing of assumptions about the error term, and model validation.

Motivation for the Study

During the seventies and eighties, interest in formal preparation for careers in business has grown rapidly. At many business schools, the number of applications for admission to the MBA programs has far exceeded the number of places available, year after year. The ratio of applicants to available places is higher than ten to one at the most prestigious schools. All schools have the challenging task of developing appropriate criteria for selecting students. Individuals charged with responsibility for selection not only have many applications to consider, they also face the task of making one overall recommendation based on a multitude of diverse pieces of information submitted by each applicant.

For example, an applicant provides the results of the Graduate Management Admissions Test (GMAT), consisting of a verbal, a quantitative, and a total score. The applicant's file also contains information on the undergraduate grade-point average (UGPA). The measurement scales used for these variables, however, are not directly comparable. For example, how is one to choose between two candidates if one has a higher GMAT score, and the other has a higher UGPA? This comparison is further complicated if the relevance of the UGPA score depends on the quality of the undergraduate institution and the undergraduate major chosen. And how is one to integrate this information with the letters of recommendation, the applicant's work experience (if any), the quality of answers to essay questions, and so forth?

In practice, admissions officers develop procedures to combine the disparate data provided by a given applicant. For example, the GMAT scores may be converted to a common scale. Once all information is converted to this common scale, a total score may be computed. Still, many subjective judgments are required for this process. For example, the maximum number of points to be allocated for the GMAT score on the common scale has to be determined. Essentially, this decision about the maximum number of points available for each variable results in the use of differential weights for each of the variables. Also, one must determine how to convert a given GMAT score to an appropriate number of points on the common scale. Further, how should one distinguish between different institutions and different undergraduate majors? Because of these subjective elements, it is very likely that two admissions officers working for the same school could obtain different evaluations of a given applicant's potential. This subjectivity is increased further if there is disagreement about the definition of potential. For example, should the quality of an applicant be measured in terms of academic abilities or management (career) potential or both?

The existence of many subjective elements in the evaluation process also makes it likely that a given admissions officer's evaluation is influenced by the quality of applicants reviewed just prior to the one under consideration. If there are no absolute standards for admission, candidates will be compared relative to each other. Thus, an applicant's chances for admission may be greater if the previous candidates are of relatively low quality. In addition, other factors such as fatigue, boredom, and other demands on the officer's time may affect the officer's judgments.

Even if admissions officers can develop a process that is entirely consistent, one has to question the appropriateness of the weights used. That is, the method of converting, say, the GMAT scores to a commmon scale may not be optimal. To investigate this, the officers would have to compare their decisions to the actual performance of the admitted applicants who enroll in the MBA program. Such comparisons, if done at all, tend to happen in a haphazard manner. For example, the instructor in an accounting class may complain to the admissions officers that some of the students are incapable of learning the material. This type of pressure may result in changes in the weights used for the different variables. That is, whatever the officers believe to be related to a capacity for accounting may increase in importance in the evaluation process.

These considerations suggest that it may be advantageous to use regression analysis as a basis for developing a model which can be used as an aid in the admissions decisions.

The advantages of using regression analysis include the following:

1. A decision is required about the definition of the criterion variable (e.g., what should the successful applicants' records or skills look like?). This decision can only be made after a discussion involving the relevant parties; for example, top administrators of the school, faculty members, and the school's graduates may be asked to participate in such discussions.
2. A set of potentially relevant predictor variables has to be identified and measured. No agreement is required on the members of this set; any viable variables on which information is available at the time of the application can be included. A discussion about what variables may be relevant, in fact, can lead to changes in the application form.
3. Once data have been gathered and a regression model is developed and accepted, this model provides the *optimal* weighting of predictor variables relevant to predicting the criterion variable chosen.
4. The model provides consistent predictions. It does not matter whether an applicant is evaluated after an exceptionally qualified or a poorly qualified applicant; the prediction of a given applicant with certain characteristics is unaffected by the order of processing applicants.

Of course, whatever model is adopted cannot be complete. Applicants may provide information that is difficult or impossible to include in the set of predictor variables. Also, the criterion variable chosen fails to consider aspects that one may like to include. For example, if the criterion variable is defined as academic performance, how should one take into consideration important aspects not captured by this criterion variable? Because of such limitations, the model can only serve as an aid in decision making. For optimal use of the model as a decision aid, the admissions officers must have complete understanding of the model and its components. If they do, they know what aspects are outside the model, and they can use their judgment as to how to modify model predictions and consider other factors not captured by the criterion variable.

Role of the Model

We argue that the proposed model is more consistent and valid with respect to integrating diverse information for the prediction of the chosen criterion variable than human judgment alone. The use of a model also allows this prediction to be made very quickly, and it should allow the admissions officers to focus more of their time on the qualitative considerations that are left out of the model.

At Stanford, applicants are evaluated both in terms of academic abilities and in terms of management potential. At the time of the study, the decision was made to develop an initial model only for predicting academic performance. Thus, the admissions officers would not only have to consider reasons for modifying the model predictions, based on unique factors excluded from the model, but also make a subjective judgment about management potential. The suggested procedure consists of two stages. First, academic performance is predicted, and the admissions officers have an opportunity to modify the predictions. Second, the officers select those who are judged to have the highest management potential from all applicants predicted to be academically viable. Thus, the model only provides information relevant to an intermediate step in the evaluation process, and it is therefore truly a decision aid.

Academic Performance

It is convenient to operationalize academic performance by the grade-point average students obtain in the MBA program. However, the program consists of distinct elements that require different abilities and skills. For example, a course in statistics requires a facility with formulas and data. On the other hand, a course in business policy emphasizes the integration of the functional areas of business and requires managerial skills. Thus, students' performance in statistics may not be strongly (positively) correlated with their performance in business policy. Of course, a grade-point average measured across all courses taken does not show how each course contributed to this average. If a student lacks the managerial skills required for courses such as business policy, the student may be able to compensate for the poor performance with a superior performance in courses such as statistics. If it is considered important for students to have adequate skills and performance in such different courses, an overall grade-point average is not a useful (measurement) of academic performance.

   Both statistics and business policy are required of all MBA students. For required courses, the performance is comparable across students, because it represents a common experience. On the other hand, students choose widely different elective courses. Thus, the grade-point average determined by required courses involves a better opportunity for explanation than is true for the elective courses. Students who perform relatively poorly on the required courses may have a tendency to choose electives where the opportunity to compensate for a poor performance is high. This suggests that it is important to have separate measures for required and elective courses.

   If we continue this argument, we may propose that all courses should be treated separately, since each course is at least somewhat different from any other course in the skills required for successful completion. However, it is not practical to treat each course separately. Also, the reliability of a grade-point average based on, say, nine courses is considerably greater than the reliability of the score in one course. If the true variability in grades is the same in each of the nine courses, then the standard error of the grade-point average (the mean) is one-third of the standard deviation for the grades in one course; that is,

$$\sigma_{\bar{X}} = \frac{\sigma}{\sqrt{n}} = \frac{\sigma}{3} \text{ if } n = 9$$

where σ is the standard deviation in grades for an individual,

   n is the number of grades (one per course), and

   $\bar{X}$ is the grade-point average.

For these reasons, academic performance in the MBA program is measured as follows. There are three criterion variables, and a separate regression analysis is carried out for each of these variables. The first criterion variable is defined as management: MGMT = the grade-point average for the required courses in finance, marketing, macroeconomics, organizational behavior, business policy, and business in a changing environment. The second criterion variable is defined as quantitative: QUANT = the grade-point average for the required courses in decision-sciences I and II (probability theory and mathematical programming), data analysis, accounting I and II, microeconomics, and computers. The third criterion variable is measured by the average performance in all elective courses.

There are also procedures available to derive such course groupings empirically. For example, we could group courses based on the magnitude of correlation between grades in each pair of courses (see Table 11.1). The higher the positive correlation, the more appropriate it is to have the respective courses together in the same group. In Table 11.1, correlations are reported for the thirteen required courses, for 296 students who graduated in 1978. The highest correlation is obtained for the two accounting courses ($r = 0.67$), and the lowest is obtained for organizational behavior and decision sciences II ($r = 0.01$). The first seven courses listed comprise QUANT, while the next six courses are used to define MGMT. An inspection of Table 11.1 suggests that the correlations between the QUANT courses (the numbers above the horizontal line) and the correlations between the MGMT courses (the numbers to the right of the vertical line) tend to be higher than the remaining correlations between one course of MGMT and one course of QUANT.

Given the choice of three separate criterion variables, we also have to develop three regression models. Nevertheless, in general we consider the same set of predictor variables for each of the three models. We do expect, however, that the influence of a given predictor variable is systematically different between these models. For example, the quantitative score on the GMAT should be strongly related to the QUANT GPA but weakly related, if at all, to MGMT GPA. Similarly, the verbal score on the GMAT may be strongly related to MGMT GPA but weakly related, if at all, to QUANT GPA. If one overall GPA criterion variable is used, the effect for a given predictor variable is, at best, a weighted average of the same predictor's effects on the three separate criterion variables.

Predictor Variables

There is a substantial collection of literature on the potential relevance of specific predictor variables. Given that the application form is designed to help admissions officers evaluate, among other things, the academic ability of each applicant, we would expect that information is available on those predictor variables included in previous empirical studies. Some of these predictor variables are directly available in numerical form and are readily available for analysis. This set includes the GMAT scores, undergraduate GPA, the amount of business experience, and age. Other variables, such

Table 11.1
Correlations between grades in required courses for the class of 1978 ($n = 296$)

Courses	Dec. sci. I	Dec. sci. II	Data anal.	Acctg. I	Acctg. II	Microecon.	Computers	Finance	Marketing	Macroecon.	Org. Behav.	Policy	Environment
Dec. Sci. I	1.00												
Dec. Sci. II	0.57	1.00											
Data Anal.	0.60	0.53	1.00										
Acctg. I	0.61	0.45	0.52	1.00									
Acctg. II	0.58	0.49	0.52	0.67	1.00								
Microecon.	0.55	0.39	0.52	0.45	0.48	1.00							
Computers	0.32	0.36	0.39	0.29	0.30	0.29	1.00						
Finance	0.29	0.27	0.32	0.39	0.41	0.24	0.20	1.00					
Marketing	0.17	0.13	0.22	0.21	0.26	0.23	0.22	0.31	1.00				
Macroecon.	0.27	0.21	0.31	0.32	0.27	0.32	0.21	0.33	0.28	1.00			
Org. Behav.	0.10	0.01	0.06	0.08	0.04	0.13	0.19	0.13	0.24	0.16	1.00		
Policy	0.09	0.06	0.15	0.06	0.09	0.09	0.17	0.16	0.11	0.15	0.28	1.00	
Environment	0.21	0.14	0.28	0.23	0.22	0.22	0.19	0.28	0.25	0.34	0.25	0.23	1.00

as undergraduate major, can be incorporated by means of indicator variables. And the quality of the undergraduate institution can be quantified by the average GMAT score of the undergraduate students graduating from the same institution. These variables are considered to have primary relevance to the prediction of academic performance.

Of secondary relevance are variables that measure applicants' achievements, leadership activities, demonstrated initiative, and so forth. Although such variables may help explain academic performance, their strength in explanatory power is expected to be small, given the inclusion of the other predictors. These variables are also more difficult to quantify. For each of these variables a rating scale is required that converts information available in the application to a numerical scale. The numerical scale needs to be well-defined in the sense that different raters should obtain the same rating of a given applicant.

The distinction between the two sets of predictor variables is important for several reasons. The first set, including GMAT scores and undergraduate grades, consists of variables that are easily accessible. These variables also tend to be quantified. The second set includes variables that are more difficult to quantify. No matter how much care is used to construct rating scales for these qualitative variables, different raters are likely to differ in the scores they assign to a given applicant. This unreliability makes it more difficult to obtain statistically significant effects for the qualitative variables. (See also the discussion of errors in predictor variables in Chapter 9.) These qualitative variables are also likely to be more relevant for predicting career performance than for predicting academic performance. Consequently, the starting model contains only those predictor variables that are easily accessible. Other variables are considered in a later stage to see if they provide a statistically significant amount of additional explanatory power. This use of the two sets of predictor variables is consistent with a greater emphasis on prediction than on understanding in model development.

The Sample

Data from the graduating classes of 1977 and 1978 were used for model development, testing, and validation. Due to the substantial number of potentially relevant predictor variables (see below), including many indicator variables, it was expected that several iterations would be used to obtain a final model. The larger this number of iterations, the greater the possible upward bias in the goodness-of-fit measures reported for the final model. For that reason, the total sample of 629 graduating students was randomly split into estimation (80 percent) and validation (20 percent) samples.

Subgroup Differences

There is frequent speculation about systematic differences between subgroups of the student body at a graduate business school. For example, female students may perform differently from male students. It is possible that such differences exist even when relevant predictor variables are held constant. Also, members of minority groups may perform differently from others. And it seems appropriate to consider the unique aspects of foreign students. In fact, some of the easily accessible data for American students is difficult or impossible to obtain for foreign students. Due to the widely varying characteristics of undergraduate education across nations, for example, there is often no undergraduate grade-point average available for foreign students.

To investigate the possibility of subgroup differences in model parameters, model development takes place based on the observations that represent American white males. This is based on the fact that they are the largest subgroup, so that a substantial sample size is available. Once a final model is accepted for this group, subgroup differences are investigated. Even if there is no statistical evidence that differences exist, it is instructive to examine the predictive validity of the model separately for each subgroup in the validation sample.

Missing Data

For two of the predictor variables, undergraduate grade-point average and a quality index of the undergraduate institution, foreign students ended up with a zero value because grade point averages are not comparable and a quality index is often not available. To compensate for this deficiency in data, an indicator variable (1 if foreign, 0 otherwise) is added to the model. In other cases, an observation was excluded entirely from the analysis if information on one or more predictor variables was missing (this is the "listwise deletion" option for handling missing data). However, in the final equations, the set of remaining predictor variables may differ between the three criterion variables. Consequently, the number of observations on which each of the three final models is based may not be the same.

To minimize the loss of observations due to missing data, other options were considered. Specifically, among American students there was a relatively small proportion of observations for which one year of undergraduate grades was missing. If the student applied for admission during the senior year of undergraduate education, for example, grades received for that year would be lacking. Such missing data could be predicted, however, based on the grades received during the first three years. The procedure for this is the following. First, identify all observations with complete data for each of the four years of undergraduate education. Second, estimate the relationship between the GPA in year four as a function of the GPAs in the first three years. Third, use the estimated relationship to predict the fourth-year GPA for the students for whom only that year's GPA is missing.

The complete set of predictor variables (listed alphabetically) used for model development is shown in Table 11.2. The variable AGESQ (squared values of age) is used, in addition to AGE, in order to allow for a nonlinear effect of age on each of the three criterion variables.

If AGE and AGESQ have positive and negative slope coefficients, respectively, then the effect of increases in age changes from positive to negative (see Chapter 6). For

Table 11.2
List of easily accessible predictor variables

ADV	= degree received as a graduate student (1 if received an advanced degree, 0 otherwise)
AGE	= age of student at time of enrollment in the MBA program (in months)
AGESQ	= the squared value for AGE
CES	= candidate excellence by school index[1]
CLASS	= year of graduation from MBA program (1 if 1977, 0 if 1978)
DUMJI	= undergraduate major area
DUMJI	= 0 for all i,

except   DUMJI1 = 1 if math., statistics or computer science
               DUMJI2 = 1 if behavioral sciences
               DUMJI3 = 1 if other sciences
               DUMJI4 = 1 if engineering
               DUMJI5 = 1 if business administration
               DUMJI6 = 1 if accounting
               DUMJI7 = 1 if economics
               DUMJI8 = 1 if liberal arts
               DUMJI9 = 1 if political science

EXPBUS	= full-time business experience in months
EXPMIL	= full-time military experience in months
EXPOTH	= full-time other experience in months
EXPPT	= part-time work experience in months
FOREIGN	= indicator for country of citizenship (1 if foreign, 0 otherwise)
GMATMS	= number of times student has taken graduate management admission test (GMAT) (0 if taken once, 1 if taken more than once)
GMATQ	= graduate management admission test score (quantitative)[2]
GMATV	= graduate management admission test score (verbal)[2]
GPA1	= undergraduate grade-point average as a freshman[3]
GPA2	= undergraduate grade-point average as a sophomore[3]
GPA3	= undergraduate grade-point average as a junior[3]
GPA4	= undergraduate grade-point average as a senior[3]
GRADWK	= number of months of study in a graduate school
GRADYR	= year of graduation from college
MARDUM	= marital status (1 if married, 0 otherwise)
MAXSAL	= maximum monthly salary (dollars) at time of application (0 if no prior experience, or if no salary data available, cf. NOSAL)
MILDUM	= indicator for military experience (1 if veteran, 0 otherwise)
NOSAL	= indicator for nonavailability of salary data (1 if no salary data, even if the applicant has had prior work experience; 0 otherwise)
SUMEMP	= summer work experience while in college in months
TOEFL	= test of English as a foreign language (total) score

[1] This index of quality of the undergraduate school is published by the Educational Testing Service. It is computed as the average Graduate Management Admission Test Score (total) of all students from the undergraduate school of the student who took the test.

[2] The most recent score was used for applicants who had taken the test more than once.

[3] On a four-point scale with D = 1, C = 2, B = 3, and A = 4.

example, the older (the more experienced) the student when enrolled in the MBA program, the more the student can usefully contribute and relate to the applied content of courses. But this positive effect may be the highest for initial increases in age, then diminish (the positive effect becomes smaller) as the applicant becomes older, and the effect may be negative for increases beyond some age. Apart from a possible nonlinear effect for age, only linear effects are accommodated for the continuous variables.

The predictor variables listed in Table 11.2 include four undergraduate grade-point averages, one for each of undergraduate education. The use of four separate measures increases the flexibility of the possible effects of undergraduate performance on the criterion variables. However, if the slope coefficients for these variables are not significantly different from each other, these variables can be combined. Similarly, it may not be necessary to have nine separate indicator variables for different undergraduate majors. To minimize the number of indicator variables for undergraduate major in the final model, schemes were developed based on expected similarities between the majors.

Model Development

The list of predictor variables, shown in Table 11.2, was reduced based on F- and t-statistics. For example, the equivalence of undergraduate majors was investigated with an F-statistic. The model was run through many iterations; after each run a reduced model was re-estimated to allow for the possibility that multicollinearity might be responsible for small t-ratios in earlier, more fully specified equations. A final model was obtained for American white male graduates, then the hypotheses that the relationships depend on gender and minority status were tested. The null hypotheses of no difference in the effects of the predictor variables between these groups could not be rejected.

Once the model was estimated for all American students, the foreign students' observations were added. As we indicated, an indicator variable to distinguish between foreign and American students was added, to compensate on the average for the zero values assigned to foreigners on the variables representing the quality of the undergraduate institution and the undergraduate grade-point averages. It was considered likely that there is some difference in the relationship for the foreigners. Consequently, the effect of the verbal GMAT score was allowed to depend on the student's citizenship. Specifically, the verbal GMAT score may have a greater impact for foreigners than for American students. Such an interaction effect is most likely to apply to the MGMT and ELECT criterion variables. In the required courses emphasizing quantitative skills, the verbal score should have only a minor impact.

The final model for each of the three criterion variables includes only those predictor variables with substantial t-values. Due to the proprietary nature of the research (some of the details are confidential), the slope coefficients cannot be shown. Instead, the standardized slope coefficients (beta weights) for the final model are shown in Table 11.3. As we have argued in Chapter 8, there are severe limitations associated with the interpretation of beta weights. Nevertheless, these results are somewhat useful in providing estimates of the explanatory power of the predictor variables.

Table 11.3
Beta weights and other statistics for final models based on the total sample

Predictor variable[2]	Criterion variable[1]		
	MGMT	**QUANT**	**ELECT**
AGE	0.20 (4.70)	—	1.59 (3.10)
AGESQ	—	—	−1.37 (−2.67)
CES	0.73 (3.89)	0.16 (1.12)	0.60 (3.16)
CLASS	—	−0.05 (−1.89)	—
DUMJ2456[2]	—	—	−0.09 (−2.32)
DUMJ289[2]	—	−0.07 (−2.17)	—
DUMJ7	0.14 (3.60)	—	—
EXPBUS	—	0.12 (3.93)	—
FOREIGN	1.31 (5.65)	0.89 (5.06)	0.96 (3.55)
FGMATV[2]	—	—	0.32 (1.90)
GMATV	0.30 (6.93)	0.13 (3.40)	0.09 (1.69)
GMATQ	—	0.54 (15.09)	0.13 (2.95)
GMATMS	—	−0.09 (−2.89)	—
GPA134[2]	0.62 (5.17)	0.70 (7.81)	—
GPA4	—	—	0.78 (6.85)
Adjusted R^2	0.183	0.532	0.173
Standard deviation of residuals[3]	0.198	0.229	0.177
Sample size	596	587	602

[1] Numbers in the main table are beta weights; t-ratios are in parentheses. Since the models were developed based on the estimation subsample of the total sample, the t-ratios should be interpreted only as crude indications of statistical significance. A blank denotes that the predictor variables do not appear in the model for the corresponding criterion variable.

[2] See Table 11.2 for definitions of these variables, except as noted below.

DUMJ2456 = if the undergraduate major is behavioral sciences, engineering, business administration, or accounting, and zero otherwise

DUMJ289 = if the undergraduate major is behavioral sciences, liberal arts, or political sciences, and zero otherwise

FGMATV = FOREIGN ∗ GMATV; i.e., FGMATV = GMATV for foreigners and zero otherwise

GPA134 = average GPA for freshman, junior, and senior years

[3] The values for MGMT, QUANT, and ELECT range from −1 to 1.

For the MGMT criterion variable, age and the quality of the undergraduate institution appear in the final model. Of the undergraduate majors, only the distinction between economics and other majors remains. That is, holding the other predictor variables in the final model constant, economics majors perform better. Of the GMAT scores, only the verbal component has a significant (positive) effect. The undergraduate grade-point average is measured for three years only. No significant effect was obtained for the sophomore year. The adjusted R^2 value is 0.183, and the standard deviation of residuals is 0.198. In interpreting this standard deviation, we note that each of the criterion variables has a theoretical range from −1 (worst) to +1 (best).

In the final QUANT equation, the undergraduate institution quality shows up with minor explanatory power and a low *t*-ratio. This predictor variable would be the first candidate for deletion. The negative coefficient for CLASS indicates that the graduating class of 1978 has somewhat lower academic performance than the 1977 graduates, on the average, holding the other predictors constant. In terms of undergraduate majors, those with a background in behavioral sciences, liberal arts, or political science compare unfavorably with the other undergraduates, holding all other predictor variables constant. The verbal GMAT score is somewhat relevant, while the qualitative score has a strong influence. There is also a penalty for students who have taken the GMAT more than once. For QUANT, the adjusted R^2 value is substantially higher (0.532) than for MGMT, even though the standard deviation of residuals (0.229) is also higher for QUANT. This result obtains, because there is substantially more variation in the QUANT variable than in MGMT.

The final ELECT equation includes both AGE and AGESQ. The high absolute beta weights (greater than one) for these variables are the result of the high degree of correlation between the two variables. In the electives, undergraduates in the behavioral sciences, engineering, business, and accounting have lower grades, relative to the others, holding other variables constant. Of the undergraduate grades, only the senior year has significant explanatory power in the final model. This equation includes an interaction effect for the verbal score of GMAT. The positive coefficient for FGMATV suggests that increases in this score have a greater effect for the performance of foreign students in elective courses than is true for American students.

The second set of predictor variables, incorporating ratings based on qualitative information supplied by the students, provided only minor increases in adjusted R^2 values. Of this set, a variable that measures the amount of work experience weighted by the level of experience, based on the number of levels the applicant was below top management, was by far the most promising. From a predictive perspective, however, this information may not add sufficiently to the models' explanatory power, given the cost of collecting it for all applicants. We consider this aspect when the predictive validity is discussed.

Testing of Assumptions

Residual plots did not suggest the existence of additional nonlinearities in the relationships. There was also no evidence of heteroscedasticity in these residual plots. Due to the cross-sectional nature of the data, the assumption of independence between the error terms cannot be investigated in the manner that is customary for time-series data. Also, for each criterion variable, the null hypothesis of normality for the error terms can not be rejected.

Although there is no evidence that the error-term assumptions are violated, it is also useful to consider the possible influence of outliers. However, the pattern of residuals does not reveal the existence of more outliers than would be expected based on chance variation. The possibility that the predictive validity is systematically lower for specific subgroups is considered below.

The use of three criterion variables for essentially the same sample of graduating students suggests the possibility that the value for the error term in one model may be related to the error value in another model. For example, one student may have positive residuals for all three equations, while another student may have three negative residuals. Such systematic patterns in the residuals may be due to relevant omitted variables which the three models have in common. The residual correlations, computed across the students whose observations were included in all three models, vary from 0.389 (QUANT and MGMT) to 0.593 (MGMT and ELECT). The ability of students to perform systematically better (or worse) on all three criterion variables than predicted may, of course, be due to factors that cannot be identified at the time the application is considered. In any event, no information was available to expand the models and reduce the positive correlation between the residuals.

Measurement Error in the Criterion Variable

Each of the three criterion variables is defined as the grade-point average for a subgroup of courses. For MGMT the number of courses is six, and for QUANT the number is seven. Due to the limited number of courses on which these averages are based, there is a considerable amount of measurement error in the criterion variables. For example, if each course that is part of QUANT is a perfect indicator of students' quantitative skills, then a given student should receive identical scores in these courses. However, the simple correlation coefficients between pairs of courses (see Table 11.1) grouped together are all substantially below one. The lower these coefficients, the lower the *maximum* R^2 value obtainable for the criterion variable. For MGMT the maximum is estimated to be 0.574, and for QUANT it is 0.854. The observed adjusted R^2 values may be compared to these maximum values to see what proportion of the maximum achievable is accounted for by these models (0.319 and 0.623, respectively). No maximum could be computed for ELECT, due to the wide variety of courses on which this variable is based.

Curtailment

The model developed to predict academic performance of applicants is based entirely on students enrolling in the MBA program. These students are selected based on certain criteria (for example, relatively high GMAT scores and undergraduate grade-point averages) and differ, therefore, systematically from other applicants. For that reason it is important to establish whether the model applies to the full range of possible values in the applicant population. Only to a limited extent, given the restricted range of values for at least some predictor variables in the enrollment sample, can this be assessed with residual plots.

This question can be considered indirectly, based on the slope coefficients. That is, does each model provide reasonable predictions for the full range of possible predictor variable values? Yet, the assessment of predictions is not straightforward if the criterion variable is also subject to restrictions. For example, a student's average score is restricted to the range $(-1, +1)$. Thus, no matter how excellent a given applicant's credentials, the prediction cannot exceed $+1$. And if all enrolled students have excellent credentials, some will receive relatively poor grades (to an extent, scores are assigned on a relative basis). These considerations make it more difficult to question the model's validity with an examination of hypothetical cases.

We can discuss the effects of curtailment (restricted variation) in predictor variables, if we asssume the criterion variables to have an unlimited range. For example, for the simple linear model

$$Y_i = \alpha + \beta X_i + u_i$$

and under the usual assumptions, we can show that the larger the amount of variation in X_i, the higher the observed t-ratio, and the higher the value of R^2. Thus, in general, a predictor variable that is not statistically significant in a multiple regression based on data for the enrolled students may be significant if the variation in the total applicant population applies. In addition, a larger amount of variation in one or more predictor variables changes the multicollinearity. Unfortunately, the impact on multicollinearity cannot be easily determined. However, by computing the ratio of the standard deviation of a predictor variable in the applicant pool over the standard deviation, for the same variable, in the enrollment population, an index of curtailment is obtained. The values for this index did not suggest that an adjustment for curtailment would materially affect the inclusion of predictor variables in the three models.

Model Validation

Given the iterations used to obtain the final model, it is likely that the goodness-of-fit statistics (R^2, standard deviation of residuals) are overstated. If our primary concern is with the models' predictive accuracy, it is critical to validate the model on new data. For this purpose, 20 percent of the sample of 1977 and 1978 graduates was randomly selected to be excluded from the data used for model development. It is also of interest to determine the predictive validity separately for subgroups of students. For example, it is likely that the models do not perform as well for foreign students as for others, due to the missing data on the quality of the undergraduate institution and the undergraduate grade-point average.

We report predictive validity results in Table 11.4. For each criterion variable, the results include s_y^2, the variance in the criterion variable across all individuals in the validation sample; MSE, the mean squared prediction error (comparable to the variance of the residuals in the estimation sample); r^2, the squared correlation coefficient between the actual and predicted values; and n, the sample size. The results are shown for four subgroups and for the total sample. Sample sizes differ somewhat across the criterion variables due to the pattern of missing data.

The results indicate that the subgroup of minorities has the highest amount of variability in all three criterion variables. Foreign students have the lowest amount of

Table 11.4
Predictive validity of academic performance models (1977–1978 holdout sample)

	American white males	American females	American minorities	Foreign students	All
Criterion variable: MGMT					
$S_Y^2(\times 100)$	5.76	4.58	6.00	4.45	5.57
MSE$(\times 100)$	5.06[1](5.02)[2]	3.13(3.69)	3.31(2.92)	4.20(4.54)	4.54(4.58)
r^2	0.12(0.15)	0.32(0.08)	0.75(0.81)	0.10(0.08)	0.18(0.17)
n	56[3](54)	23[3](22)	17[3](15)	19(18)	115[3](110)
Criterion variable: QUANT					
$S_Y^2(\times 100)$	11.09	10.05	12.96	8.35	11.09
MSE$(\times 100)$	5.52(5.57)	5.57(6.00)	4.67(4.45)	9.80(9.30)	6.05(6.05)
r^2	0.50(0.44)	0.36(0.32)	0.68(0.70)	0.01(0.03)	0.45(0.42)
n	56(54)	23(22)	17(15)	18(17)	114(109)
Criterion variable: ELECT					
$S_Y^2(\times 100)$	3.72	3.65	4.37	3.76	3.76
MSE$(\times 100)$	3.13(3.03)	3.53(3.84)	3.42(3.20)	4.16(4.16)	3.39(3.35)
r^2	0.16(0.18)	0.07(0.01)	0.25(0.22)	0.04(0.06)	0.10(0.09)
n	57(55)	24(23)	17(15)	19(18)	117(112)

[1] This number represents the result based on the variables included in Table 11.3.

[2] The number in parentheses represents the results obtained by adding selected rating-scale variables to the equation.

[3] The subgroups are not mutually exclusive, because American women belonging to a minority are counted in two groups. The categories are also not collectively exhaustive, because individuals who did not identify minority status are not included in any subgroup but are included in the total.

variability for the first two criterion variables. For the MGMT variable, all subgroups have a mean squared error value below the corresponding variance in the criterion variable. However, the foreign-student sample shows only a small difference. This is also indicated by the low r^2 value for foreign students. For the QUANT variable, there is a substantial difference between the values for s_Y^2 and MSE, except for foreign students. Indeed the model for foreign students results in a greater error variance than exists in the raw data. This implies that the model is worse for predictive purposes than not using a model at all! The same comparison for the ELECT equation also indicates that using the model results in more error than using no model.

The last column in Table 11.4 shows r^2 values that are below the adjusted R^2 values reported for the estimation model in Table 11.2. The difference (for example, $\bar{R}^2 = 0.532$ and $r^2 = 0.45$) suggests the magnitude of overfitting that is due to the iterations in model development.

The values in parentheses included in Table 11.4 are obtained by determining the models' predictive validities after the most promising rating variables are added. For QUANT and ELECT, two variables are added, one to capture the work experience obtained by applicants weighted by the level of experience, and the other to quantify the overall quality of the way the applicants present the information. For MGMT, in addition to these variables, a set of indicator variables was added to capture demonstrated creativity in artistic work, publications, inventions, and business ventures. In general, the predictive validity is somewhat lower when the augmented model is used, if we compare the r^2 values for the sample of all applicants. Interestingly, for the female subgroup the augmented model is worse for all criterion variables, while for the minorities the augmented model tends to do better. The overall pattern is such that the gain is minimal at best. It appears appropriate, therefore, to conclude that the initial model, with the variables reported in Table 11.2, should be considered for use.

The predictive validity results shown so far suggest that the models are somewhat useful for predictive purposes; actual use, of course, would involve future classes of students. The model developed on the graduating classes of 1977 and 1978 could be used to predict the academic performance of applicants to the class of 1979, but it might also be applied to classes later than that of 1979. Importantly, however, the content of some courses and the identity of some instructors change from year to year. Also, the characteristics of the student body may change systematically over time. Thus, it is important to determine the models' predictive validity for a subsequent class of students, and to reestimate parameters periodically using more recent data.

We show the models' predictive validity for the class of 1979 next to the corresponding measures for the 1977-78 validation and estimation samples in Table 11.5.

Table 11.5

Predictive validity for 1977–78 holdout sample and 1979 sample (with comparison to 1977–78 estimation sample)

	1977–78 holdout sample	1979 validation sample	1977–78 estimation sample
Criterion variable: MGMT			
$S_Y^2(\times 100)$	5.57	4.67	4.59
MSE($\times 100$)	4.54	3.84	3.75
r^2	0.18	0.18	0.18
n	115	222[1]	466
Criterion variable: QUANT			
$S_Y^2(\times 100)$	11.09	9.73	11.29
MSE($\times 100$)	6.05	5.38	5.10
r^2	0.45	0.48	0.55
n	114	222[1]	459
Criterion variable: ELECT			
$S_Y^2(\times 100)$	3.76	3.31	3.84
MSE	3.39	2.96	3.14
r^2	0.10	0.15	0.18
n	117	220[1]	470

[1] The sample sizes are substantially below the number of students who enrolled for this class. This is due to the fact that data on one or more predictor variables were missing for many observations.

The results suggest that the performance for the 1979 class is at least as good as for the 1977-78 validation sample. Thus, the model is as applicable for the 1979 class as it is for the classes used for model development. Nevertheless, it is important to monitor the course content (especially of the required courses) to see if there are reasons to expect a systematic change in model parameters.

An alternative method to investigate the possibility of differences in the parameters between the 1977-78 and 1979 classes is to use a test of the null hypothesis that the parameters for a given model are equal. An incremental R^2 test can be used for this purpose. The question is whether the R^2 obtained when 1979 and 1977-78 are allowed to have unique parameters is significantly greater than the R^2 resulting when the parameters are forced to be the same. The null hypothesis of no difference could not be rejected at the five-percent level of significance for any of the three models.

Implementation

The models developed can be used to predict MGMT, QUANT, and ELECT scores for each applicant separately. If the objective is to use the models to identify academically viable applicants, applicants must have a respectable predicted score on both MGMT and QUANT (the two sets of required courses). On the other hand, if we want to obtain a predicted average score for all required courses, we can take a weighted average of the predicted MGMT and QUANT scores (the weights reflecting the total number of credits for the courses). Similarly, a predicted weighted average score can be computed across all courses.

Assuming that students have to be competent in both MGMT and QUANT, and that failure is unlikely for the average performance in ELECT, academic viability can be measured by the *probability* that a given applicant fails the required courses. Failure is defined as obtaining a grade-point average below zero in either MGMT, or QUANT, or both. This measure is an easily interpretable indication of the academic risk the admissions officers take if the applicant is accepted. The probability is obtained by using the predicted MGMT and QUANT scores, the estimated standard errors of the predictions, and the correlation between the prediction errors. Given our inability to reject the assumption of normally distributed errors (see the section on testing assumptions), the normal distribution is used to obtain the probability of failure.

Once these calculations are completed, it is important to consider the impact of the specific probability used to screen applicants. For example, if all applicants with a predicted probability of failure equal to 0.1 or higher are screened out, the remaining applicants may be too small in number for the admission director to consider other factors. Furthermore, the remaining set may be lacking in diversity of background and experience. Thus, it is useful to determine the size of the application population remaining, as well as its characteristics. Of course, the actual use of the predictions is different, in the sense that admissions officers have an opportunity to modify the prediction for an applicant if that applicant has exceptional characteristics not considered by the model.

To determine the sensitivity of the size of the remaining population of applicants to different probability levels, results were obtained for several plausible probabilities. As an example, an initial population of 4,955 American applicants was reduced to 3,118, based on a 0.40 probability of failure. The reduced population has an improved average-undergraduate GPA (3.30 versus 3.17) and higher verbal (38.3 versus 35.6) and quantitative (38.2 versus 34.2) GMAT scores. The reduced population also has higher average scores on selected qualitative variables, such as the extent to which goals are expressed clearly, the articulation of the relevance of an MBA degree, and the overall presentation of the case for admission. However, the remaining population is somewhat younger and less experienced.

Nevertheless, there is a great deal of flexibility left in choosing those applicants with the highest career potential (in the admissions officers' judgments). For example, to enroll a class of 310 students, less than 450 have to be admitted. Using a 5,200 applicant pool, about 3,300 are expected to be academically viable, based on a 0.40 probability-of-failure screen. Thus, there is ample opportunity to select, for example, the most experienced applicants from the large pool of academically viable applicants.

Model Updating

If information on the easily accessible predictor variables is routinely available on a year-to-year basis, it is easy to re-estimate the parameters of the models to predict MGMT, QUANT, and ELECT. However, the need for updating is unclear, and no evidence was found regarding a difference between the class of 1979 and the classes of 1977 and 1978. Updating is important if there are systematic changes in the set of required courses, course content, or the characteristics of the student body. It is also appropriate to consider whether the composition of the MGMT and QUANT variables should be changed over time. In general, it appears appropriate to consider updating the coefficients every other year or so. In addition, the inclusion of predictor variables and the composition of the criterion variables need to be reconsidered every five years or so.

SUMMARY

In this chapter we have presented a detailed discussion of a regression application. The motivation for the application was to improve the process of making admissions decisions for an MBA program. Using a regression model to predict the academic performance of applicants should be beneficial in three respects:

1. Model predictions are consistent.
2. Model predictions are valid, in the sense that the slope coefficients are based on the relationship between academic performance and applicant characteristics.
3. Admissions officers can concentrate their efforts on making judgments about the

management potential of academically viable applicants, instead of having to make a judgment about both academic and management potential.

Although the use of regression models should aid the quality of admissions officers' judgments, it is important to stress that the officers also need to understand the model components. For example, they have an opportunity to modify model predictions, to accommodate unusual applicant characteristics not considered by the model. However, for such modifications to make sense, the admissions officers have to understand what the models capture. Thus, the model does not replace the admissions officers but serves as an aid to the decision-making process. Regression analysis does a superior job in combining information from different variables measured on noncomparable scales. In addition, the use of a model of this task makes it possibe for the admissions officers to focus their attention almost exclusively on predicting career potential for those applicants that are academically viable.

In terms of the regression analysis, we have discussed the measurement of academic performance and argued the benefits of having three separate criterion variables. The list of potentially relevant predictor variables was shown to be very long. This made it clear that a number of iterations would be required to obtain a final model. This process would also be likely to result in biased goodness-of-fit measures for the final model. For that reason, a portion of the sample was not used in model development, in order to be able to obtain a better indication of the true goodness of fit.

We also discussed the investigation of subgroup differences. Model development took place on the largest subgroup, white American males, and the possibility of subgroup differences in parameters was tested after a final model was obtained. Given a lack of evidence of systematic differences between the subgroups, it was still of interest to see if the models' predictive validities differed between the subgroups. We also discussed how and why the treatment of missing data varied dependent upon the nature and extent of the missing data.

Tests of the assumptions about the error term did not reveal shortcomings in the model specifications. Special considerations relevant to the study include the measurement error in the criterion variables, which reduces the maximum possible r^2 value.

The model validation showed how much the goodness of fit reported for the estimation model might be biased. We also saw that the predictive validity for foreign students was very poor, probably due to missing data on two key variables. Interestingly, the parameters appeared to be stable when a later third class of students was compared with the two classes used for model development.

Finally, we discussed some issues relevant to model use (implementation). Of particular importance is the conversion of regression results to measures that the ultimate users can easily relate to. For example, it is likely that admissions officers will have less difficulty with a predicted probability of failure than with one (or more) predicted grade-point averages and the uncertainty (estimated standard error) of these predictions. In this last section, we also presented reasons for updating the models. The application discussed in this chapter does not consider all possible issues that may be relevant to a particular study, but it certainly shows how a comprehensive analysis can be completed.

ENDNOTE

1. Srinivasan. V, Dick R. Wittink, and Bruce M. Zweig, "Decision Aids for MBA Program Admissions: Predicting Academic Performance," Research Paper no. 610, Graduate School of Business, Stanford University, April 1981, revised July 1981.

REFERENCE

Srinivasan, V., Dick R. Wittink, and Bruce M. Zweig. "Decision Aids for MBA Program Admissions: Predicting Academic Performance." Research Paper no. 610, Graduate School of Business, Stanford University, April 1981, revised July 1981.

Appendix of Statistical Tables

Table I
Normal probabilities: Areas of the standard normal distribution

The values in the body of the table are the areas between the mean and the value of Z.

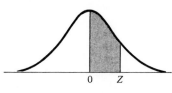

Z	.00	.01	.02	.03	.04	.05	.06	.07	.08	.09
.00	.0000	.0040	.0080	.0120	.0160	.0199	.0239	.0279	.0319	.0359
.10	.0398	.0438	.0478	.0517	.0557	.0596	.0636	.0675	.0714	.0753
.20	.0793	.0832	.0871	.0910	.0948	.0987	.1026	.1064	.1103	.1141
.30	.1179	.1217	.1255	.1293	.1331	.1368	.1406	.1443	.1480	.1517
.40	.1554	.1591	.1628	.1664	.1700	.1736	.1772	.1808	.1844	.1879
.50	.1915	.1950	.1985	.2019	.2054	.2088	.2123	.2157	.2190	.2224
.60	.2257	.2291	.2324	.2357	.2389	.2422	.2454	.2486	.2517	.2549
.70	.2580	.2611	.2642	.2673	.2703	.2734	.2764	.2793	.2823	.2852
.80	.2881	.2910	.2939	.2967	.2995	.3023	.3051	.3078	.3106	.3133
.90	.3159	.3186	.3212	.3238	.3264	.3289	.3315	.3340	.3365	.3389
1.00	.3413	.3438	.3461	.3485	.3508	.3531	.3554	.3577	.3599	.3621
1.10	.3643	.3665	.3686	.3708	.3729	.3749	.3770	.3790	.3810	.3830
1.20	.3849	.3869	.3888	.3907	.3925	.3944	.3962	.3980	.3997	.4015
1.30	.4032	.4049	.4066	.4082	.4099	.4115	.4131	.4147	.4162	.4177
1.40	.4192	.4207	.4222	.4236	.4251	.4265	.4279	.4292	.4306	.4319
1.50	.4332	.4345	.4357	.4370	.4382	.4394	.4406	.4418	.4429	.4441
1.60	.4452	.4463	.4474	.4484	.4495	.4505	.4515	.4525	.4535	.4545
1.70	.4554	.4564	.4573	.4582	.4591	.4599	.4608	.4616	.4625	.4633
1.80	.4641	.4649	.4656	.4664	.4671	.4678	.4686	.4693	.4699	.4706
1.90	.4713	.4719	.4726	.4732	.4738	.4744	.4750	.4756	.4761	.4767
2.00	.4772	.4778	.4783	.4788	.4793	.4798	.4803	.4808	.4812	.4817
2.10	.4821	.4826	.4830	.4834	.4838	.4842	.4846	.4850	.4854	.4857
2.20	.4861	.4864	.4868	.4871	.4875	.4878	.4881	.4884	.4887	.4890
2.30	.4893	.4896	.4898	.4901	.4904	.4906	.4909	.4911	.4913	.4916
2.40	.4918	.4920	.4922	.4925	.4927	.4929	.4931	.4932	.4934	.4936
2.50	.4938	.4940	.4941	.4943	.4945	.4946	.4948	.4949	.4951	.4952
2.60	.4953	.4955	.4956	.4957	.4959	.4960	.4961	.4962	.4963	.4964
2.70	.4965	.4966	.4967	.4968	.4969	.4970	.4971	.4972	.4973	.4974
2.80	.4974	.4975	.4976	.4977	.4977	.4978	.4979	.4979	.4980	.4981
2.90	.4981	.4982	.4982	.4983	.4984	.4984	.4985	.4985	.4986	.4986
3.00	.4987	.4987	.4987	.4988	.4988	.4989	.4989	.4989	.4990	.4990
3.10	.4990	.4991	.4991	.4991	.4992	.4992	.4992	.4992	.4993	.4993
3.20	.4993	.4993	.4994	.4994	.4994	.4994	.4994	.4995	.4995	.4995
3.30	.4995	.4995	.4995	.4996	.4996	.4996	.4996	.4996	.4996	.4997
3.40	.4997	.4997	.4997	.4997	.4997	.4997	.4997	.4997	.4997	.4998
3.50	.4998	.4998	.4998	.4998	.4998	.4998	.4998	.4998	.4998	.4998
3.60	.4998	.4998	.4999	.4999	.4999	.4999	.4999	.4999	.4999	.4999
3.70	.4999	.4999	.4999	.4999	.4999	.4999	.4999	.4999	.4999	.4999
3.80	.4999	.4999	.4999	.4999	.4999	.4999	.4999	.4999	.4999	.4999

Example: If we want to find the area under the standard normal curve between Z = 0 and Z = 1.96, we find the Z = 1.90 row and .06 column (for Z = 1.90 + .06 = 1.96) and read .4750 at the intersection.

Table II
Values of *t* for given probability levels

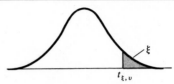

Degrees of Freedom	Probability ξ of a Larger Value				
v	.1	.05	.025	.01	.005
1	3.078	6.314	12.706	31.821	63.657
2	1.886	2.920	4.303	6.965	9.925
3	1.638	2.353	3.182	4.541	5.841
4	1.533	2.132	2.776	3.747	4.604
5	1.476	2.015	2.571	3.365	4.032
6	1.440	1.943	2.447	3.143	3.707
7	1.415	1.895	2.365	2.998	3.499
8	1.397	1.860	2.306	2.896	3.355
9	1.383	1.833	2.262	2.821	3.250
10	1.372	1.812	2.228	2.764	3.169
11	1.363	1.796	2.201	2.718	3.106
12	1.356	1.782	2.179	2.681	3.055
13	1.350	1.771	2.160	2.650	3.012
14	1.345	1.761	2.145	2.624	2.977
15	1.341	1.753	2.131	2.602	2.947
16	1.337	1.746	2.120	2.583	2.921
17	1.333	1.740	2.110	2.567	2.898
18	1.330	1.734	2.101	2.552	2.878
19	1.328	1.729	2.093	2.539	2.861
20	1.325	1.725	2.086	2.528	2.845
21	1.323	1.721	2.080	2.518	2.831
22	1.321	1.717	2.074	2.508	2.819
23	1.319	1.714	2.069	2.500	2.807
24	1.318	1.711	2.064	2.492	2.797
25	1.316	1.708	2.060	2.485	2.787
26	1.315	1.706	2.056	2.479	2.779
27	1.314	1.703	2.052	2.473	2.771
28	1.313	1.701	2.048	2.467	2.763
29	1.311	1.699	2.045	2.462	2.756
30	1.310	1.697	2.042	2.457	2.750
40	1.303	1.684	2.021	2.423	2.704
60	1.296	1.671	2.000	2.390	2.660
120	1.290	1.661	1.984	2.358	2.626
∞	1.282	1.645	1.960	2.326	2.576

Source: Abridged from Table III of Fisher and Yates: *Statistical Tables for Biological, Agricultural and Medical Research*, published by Longman Group Ltd., London (previously published by Oliver & Boyd, Edinburgh), by permission of the authors and publishers.

Example: The probability is .05 of obtaining a *t*-value in excess of 1.725 when the degrees of freedom are 20 (under the null hypothesis).

Table III
Percentage points of the F distribution

F Distribution: .05 Points

ν_2 \ ν_1	1	2	3	4	5	6	7	8	9
1	161.45	199.50	215.71	224.58	230.16	233.99	236.77	238.88	240.54
2	18.513	19.000	19.164	19.247	19.296	19.330	19.353	19.371	19.385
3	10.128	9.5521	9.2766	9.1172	9.0135	8.9406	8.8868	8.8452	8.8123
4	7.7086	6.9443	6.5914	6.3883	6.2560	6.1631	6.0942	6.0410	5.9988
5	6.6079	5.7861	5.4095	5.1922	5.0503	4.9503	4.8759	4.8183	4.7725
6	5.9874	5.1433	4.7571	4.5337	4.3874	4.2839	4.2066	4.1468	4.0990
7	5.5914	4.7374	4.3468	4.1203	3.9715	3.8660	3.7870	3.7257	3.6767
8	5.3177	4.4590	4.0662	3.8378	3.6875	3.5806	3.5005	3.4381	3.3881
9	5.1174	4.2565	3.8626	3.6331	3.4817	3.3738	3.2927	3.2296	3.1789
10	4.9646	4.1028	3.7083	3.4780	3.3258	3.2172	3.1355	3.0717	3.0204
11	4.8443	3.9823	3.5874	3.3567	3.2039	3.0946	3.0123	2.9480	2.8962
12	4.7472	3.8853	3.4903	3.2592	3.1059	2.9961	2.9134	2.8486	2.7964
13	4.6672	3.8056	3.4105	3.1791	3.0254	2.9153	2.8321	2.7669	2.7144
14	4.6001	3.7389	3.3439	3.1122	2.9582	2.8477	2.7642	2.6987	2.6458
15	4.5431	3.6823	3.2874	3.0556	2.9013	2.7905	2.7066	2.6408	2.5876
16	4.4940	3.6337	3.2389	3.0069	2.8524	2.7413	2.6572	2.5911	2.5377
17	4.4513	3.5915	3.1968	2.9647	2.8100	2.6987	2.6143	2.5480	2.4943
18	4.4139	3.5546	3.1599	2.9277	2.7729	2.6613	2.5767	2.5102	2.4563
19	4.3808	3.5219	3.1274	2.8951	2.7401	2.6283	2.5435	2.4768	2.4227
20	4.3513	3.4928	3.0984	2.8661	2.7109	2.5990	2.5140	2.4471	2.3928
21	4.3248	3.4668	3.0725	2.8401	2.6848	2.5757	2.4876	2.4205	2.3661
22	4.3009	3.4434	3.0491	2.8167	2.6613	2.5491	2.4638	2.3965	2.3419
23	4.2793	3.4221	3.0280	2.7955	2.6400	2.5277	2.4422	2.3748	2.3201
24	4.2597	3.4028	3.0088	2.7763	2.6207	2.5082	2.4226	2.3551	2.3002
25	4.2417	3.3852	2.9912	2.7587	2.6030	2.4904	2.4047	2.3371	2.2821
26	4.2252	3.3690	2.9751	2.7426	2.5868	2.4741	2.3883	2.3205	2.2655
27	4.2100	3.3541	2.9604	2.7278	2.5719	2.4591	2.3732	2.3053	2.2501
28	4.1960	3.3404	2.9467	2.7141	2.5581	2.4453	2.3593	2.2913	2.2360
29	4.1830	3.3277	2.9340	2.7014	2.5454	2.4324	2.3463	2.2782	2.2229
30	4.1709	3.3158	2.9223	2.6896	2.5336	2.4205	2.3343	2.2662	2.2107
40	4.0848	3.2317	2.8387	2.6060	2.4495	2.3359	2.2490	2.1802	2.1240
60	4.0012	3.1504	2.7581	2.5252	2.3683	2.2540	2.1665	2.0970	2.0401
120	3.9201	3.0718	2.6802	2.4472	2.2900	2.1750	2.0867	2.0164	1.9588
∞	3.8415	2.9957	2.6049	2.3719	2.2141	2.0986	2.0096	1.9384	1.8799

Example: The probability is .05 of obtaining an F-value in excess of 2.5252 when the degrees of freedom are $\nu_1 = 4$ in the numerator, and $\nu_2 = 60$ in the denominator (under the null hypothesis).

(Continued)

Table III   (*Continued*)

ν_1 / ν_2	10	12	15	20	24	30	40	60	120	∞
1	241.88	243.91	245.95	248.01	249.05	250.09	251.14	252.20	253.25	254.32
2	19.396	19.413	19.429	19.446	19.454	19.462	19.471	19.479	19.487	19.496
3	8.7855	8.7446	8.7029	8.6602	8.6385	8.6166	8.5944	8.5720	8.5494	8.5265
4	5.9644	5.9117	5.8578	5.8025	5.7744	5.7459	5.7170	5.6878	5.6581	5.6281
5	4.7351	4.6777	4.6188	4.5581	4.5272	4.4957	4.4638	4.4314	4.3984	4.3650
6	4.0600	3.9999	3.9381	3.8742	3.8415	3.8082	3.7743	3.7398	3.7047	3.6688
7	3.6365	3.5747	3.5108	3.4445	3.4105	3.3758	3.3404	3.3043	3.2674	3.2298
8	3.3472	3.2840	3.2184	3.1503	3.1152	3.0794	3.0428	3.0053	2.9669	2.9276
9	3.1373	3.0729	3.0061	2.9365	2.9005	2.8637	2.8259	2.7872	2.7475	2.7067
10	2.9782	2.9130	2.8450	2.7740	2.7372	2.6996	2.6609	2.6211	2.5801	2.5379
11	2.8536	2.7876	2.7186	2.6464	2.6090	2.5705	2.5309	2.4901	2.4480	2.4045
12	2.7534	2.6866	2.6169	2.5436	2.5055	2.4663	2.4259	2.3842	2.3410	2.2962
13	2.6710	2.6037	2.5331	2.4589	2.4202	2.3803	2.3392	2.2966	2.2524	2.2064
14	2.6021	2.5342	2.4630	2.3879	2.3487	2.3082	2.2664	2.2230	2.1778	2.1307
15	2.5437	2.4753	2.4035	2.3275	2.2878	2.2468	2.2043	2.1601	2.1141	2.0658
16	2.4935	2.4247	2.3522	2.2756	2.2354	2.1938	2.1507	2.1058	2.0589	2.0096
17	2.4499	2.3807	2.3077	2.2304	2.1898	2.1477	2.1040	2.0584	2.0107	1.9604
18	2.4117	2.3421	2.2686	2.1906	2.1497	2.1071	2.0629	2.0166	1.9681	1.9168
19	2.3779	2.3080	2.2341	2.1555	2.1141	2.0712	2.0264	1.9796	1.9302	1.8780
20	2.3479	2.2776	2.2033	2.1242	2.0825	2.0391	1.9938	1.9464	1.8963	1.8432
21	2.3210	2.2504	2.1757	2.0960	2.0540	2.0102	1.9645	1.9165	1.8657	1.8117
22	2.2967	2.2258	2.1508	2.0707	2.0283	1.9842	1.9380	1.8895	1.8380	1.7831
23	2.2747	2.2036	2.1282	2.0476	2.0050	1.9605	1.9139	1.8649	1.8128	1.7570
24	2.2547	2.1834	2.1077	2.0267	1.9838	1.9390	1.8920	1.8424	1.7897	1.7331
25	2.2365	2.1649	2.0889	2.0075	1.9643	1.9192	1.8718	1.8217	1.7684	1.7110
26	2.2197	2.1479	2.0716	1.9898	1.9464	1.9010	1.8533	1.8027	1.7488	1.6906
27	2.2043	2.1323	2.0558	1.9736	1.9299	1.8842	1.8361	1.7851	1.7307	1.6717
28	2.1900	2.1179	2.0411	1.9586	1.9147	1.8687	1.8203	1.7689	1.7138	1.6541
29	2.1768	2.1045	2.0275	1.9446	1.9005	1.8543	1.8055	1.7537	1.6981	1.6377
30	2.1646	2.0921	2.0148	1.9317	1.8874	1.8409	1.7918	1.7396	1.6835	1.6223
40	2.0772	2.0035	1.9245	1.8389	1.7929	1.7444	1.6928	1.6373	1.5766	1.5089
60	1.9926	1.9174	1.8364	1.7480	1.7001	1.6491	1.5943	1.5343	1.4673	1.3893
120	1.9105	1.8337	1.7505	1.6587	1.6084	1.5543	1.4952	1.4290	1.3519	1.2539
∞	1.8307	1.7522	1.6664	1.5705	1.5173	1.4591	1.3940	1.3180	1.2214	1.0000

F Distribution: .05 Points

Table III   (*Continued*)

ν_1 / ν_2	1	2	3	4	5	6	7	8	9
1	647.79	799.50	864.16	899.58	921.85	937.11	948.22	956.66	963.28
2	38.506	39.000	39.165	39.248	29.298	39.331	39.355	39.373	39.387
3	17.443	16.044	15.439	15.101	14.885	14.735	14.624	14.540	14.473
4	12.218	10.649	9.9792	9.6045	9.3645	9.1973	9.0741	8.9796	8.9047
5	10.007	8.4336	7.7636	7.3879	7.1464	6.9777	6.8531	6.7572	6.6810
6	8.8131	7.2598	6.5988	6.2272	5.9876	5.8197	5.6955	5.5996	5.5234
7	8.0727	6.5415	5.8898	5.5226	5.2852	5.1186	4.9949	4.8994	4.8232
8	7.5709	6.0595	5.4160	5.0526	4.8173	4.6517	4.5286	4.4332	4.3572
9	7.2093	5.7147	5.0781	4.7181	4.4844	4.3197	4.1971	4.1020	4.0260
10	6.9367	5.4564	4.8256	4.4683	4.2361	4.0721	3.9498	3.8549	3.7790
11	6.7241	5.2559	4.6300	4.2751	4.0440	3.8807	3.7586	3.6638	3.5879
12	6.5538	5.0959	4.4742	4.1212	3.8911	3.7283	3.6065	3.5118	3.4358
13	6.4143	4.9653	4.3472	3.9959	3.7667	3.6043	3.4827	3.3880	3.3120
14	6.2979	4.8567	4.2417	3.8919	3.6634	3.5014	3.3799	3.2853	3.2093
15	6.1995	4.7650	4.1528	3.8043	3.5764	3.4147	3.2934	3.1987	3.1227
16	6.1151	4.6867	4.0768	3.7294	3.5021	3.3406	3.2194	3.1248	3.0488
17	6.0420	4.6189	4.0112	3.6648	3.4379	3.2767	3.1556	3.0610	2.9849
18	5.9781	4.5597	3.9539	3.6083	3.3820	3.2209	3.0999	3.0053	2.9291
19	5.9216	4.5075	3.9034	3.5587	3.3327	3.1718	3.0509	2.9563	2.8800
20	5.8715	4.4613	3.8587	3.5147	3.2891	3.1283	3.0074	2.9128	2.8365
21	5.8266	4.4199	3.8188	3.4754	3.2501	3.0895	2.9686	2.8740	2.7977
22	5.7863	4.3828	3.7829	3.4401	3.2151	3.0546	2.9338	2.8392	2.7628
23	5.7498	4.3492	3.7505	3.4083	3.1835	3.0232	2.9024	2.8077	2.7313
24	5.7167	4.3187	3.7211	3.3794	3.1548	2.9946	2.8738	2.7791	2.7027
25	5.6864	4.2909	3.6943	3.3530	3.1287	2.9685	2.8478	2.7531	2.6766
26	5.6586	4.2655	3.6697	3.3289	3.1048	2.9447	2.8240	2.7293	2.6528
27	5.6331	4.2421	3.6472	3.3067	3.0828	2.9228	2.8021	2.7074	2.6309
28	5.6096	4.2205	3.6264	3.2863	3.0625	2.9027	2.7820	2.6872	2.6106
29	5.5878	4.2006	3.6072	3.2674	3.0438	2.8840	2.7633	2.6686	2.5919
30	5.5675	4.1821	3.5894	3.2499	3.0265	2.8667	2.7460	2.6513	2.5746
40	5.4239	4.0510	3.4633	3.1261	2.9037	2.7444	2.6238	2.5289	2.4519
60	5.2857	3.9253	3.3425	3.0077	2.7863	2.6274	2.5068	2.4117	2.3344
120	5.1524	3.8046	3.2270	2.8943	2.6740	2.5154	2.3948	2.2994	2.2217
∞	5.0239	3.6889	3.1161	2.7858	2.5665	2.4082	2.2875	2.1918	2.1136

F Distribution: .025 Points

(*Continued*)

Table III (Continued)

ν_2 \ ν_1	10	12	15	20	24	30	40	60	120	∞
					F Distribution: .025 Points					
1	968.63	976.71	984.87	993.10	997.25	1001.4	1005.6	1009.8	1014.0	1018.3
2	39.398	39.415	39.431	39.448	39.456	39.465	39.473	39.481	39.490	39.498
3	14.419	14.337	14.253	14.167	14.124	14.081	14.037	13.992	13.947	13.902
4	8.8439	8.7512	8.6565	8.5599	8.5109	8.4613	8.4111	8.3604	8.3092	8.2573
5	6.6192	6.5246	6.4277	6.3285	6.2780	6.2269	6.1751	6.1225	6.0693	6.0153
6	5.4613	5.3662	5.2687	5.1684	5.1172	5.0652	5.0125	4.9589	4.9045	4.8491
7	4.7611	4.6658	4.5678	4.4667	4.4150	4.3624	4.3089	4.2544	4.1989	4.1423
8	4.2951	4.1997	4.1012	3.9995	3.9472	3.8940	3.8398	3.7844	3.7279	3.6702
9	3.9639	3.8682	3.7694	3.6669	3.6142	3.5604	3.5055	3.4493	3.3918	3.3329
10	3.7168	3.6209	3.5217	3.4186	3.3654	3.3110	3.2554	3.1984	3.1399	3.0798
11	3.5257	3.4296	3.3299	3.2261	3.1725	3.1176	3.0613	3.0035	2.9441	2.8828
12	3.3736	3.2773	3.1772	3.0728	3.0187	2.9633	2.9063	2.8478	2.7874	2.7249
13	3.2497	3.1532	3.0527	2.9477	2.8932	2.8373	2.7797	2.7204	2.6590	2.5955
14	3.1469	3.0501	2.9493	2.8437	2.7888	2.7324	2.6742	2.6142	2.5519	2.4872
15	3.0602	2.9633	2.8621	2.7559	2.7006	2.6437	2.5850	2.5242	2.4611	2.3953
16	2.9862	2.8890	2.7875	2.6808	2.6252	2.5678	2.5085	2.4471	2.3831	2.3163
17	2.9222	2.8249	2.7230	2.6158	2.5598	2.5021	2.4422	2.3801	2.3153	2.2474
18	2.8664	2.7689	2.6667	2.5590	2.5027	2.4445	2.3842	2.3214	2.2558	2.1869
19	2.8173	2.7196	2.6171	2.5089	2.4523	2.3937	2.3329	2.2695	2.2032	2.1333
20	2.7737	2.6758	2.5731	2.4645	2.4076	2.3486	2.2873	2.2234	2.1562	2.0853
21	2.7348	2.6368	2.5338	2.4247	2.3675	2.3082	2.2465	2.1819	2.1141	2.0422
22	2.6998	2.6017	2.4984	2.3890	2.3315	2.2718	2.2097	2.1446	2.0760	2.0032
23	2.6682	2.5699	2.4665	2.3567	2.2989	2.2389	2.1763	2.1107	2.0415	1.9677
24	2.6396	2.5412	2.4374	2.3273	2.2693	2.2090	2.1460	2.0799	2.0099	1.9353
25	2.6135	2.5149	2.4110	2.3005	2.2422	2.1816	2.1183	2.0517	1.9811	1.9055
26	2.5895	2.4909	2.3867	2.2759	2.2174	2.1565	2.0928	2.0257	1.9545	1.8781
27	2.5676	2.4688	2.3644	2.2533	2.1946	2.1334	2.0693	2.0018	1.9299	1.8527
28	2.5473	2.4484	2.3438	2.2324	2.1735	2.1121	2.0477	1.9796	1.9072	1.8291
29	2.5286	2.4295	2.3248	2.2131	2.1540	2.0923	2.0276	1.9591	1.8861	1.8072
30	2.5112	2.4120	2.3072	2.1952	2.1359	2.0739	2.0089	1.9400	1.8664	1.7867
40	2.3882	2.2882	2.1819	2.0677	2.0069	1.9429	1.8752	1.8028	1.7242	1.6371
60	2.2702	2.1692	2.0613	1.9445	1.8817	1.8152	1.7440	1.6668	1.5810	1.4822
120	2.1570	2.0548	1.9450	1.8249	1.7597	1.6899	1.6141	1.5299	1.4327	1.3104
∞	2.0483	1.9447	1.8326	1.7085	1.6402	1.5660	1.4835	1.3883	1.2684	1.0000

Table III   (*Continued*)

					F Distribution: .01 Points				
ν_1 ν_2	1	2	3	4	5	6	7	8	9
1	4052.2	4999.5	5403.3	5624.6	5763.7	5859.0	5928.3	5981.6	6022.5
2	98.503	99.000	99.166	99.249	99.299	99.332	99.356	99.374	99.388
3	34.116	30.817	29.457	28.710	28.237	27.911	27.672	27.489	27.345
4	21.198	18.000	16.694	15.977	15.522	15.207	14.976	14.799	14.659
5	16.258	13.274	12.060	11.392	10.967	10.672	10.456	10.289	10.158
6	13.745	10.925	9.7795	9.1483	8.7459	8.4661	8.2600	8.1016	7.9761
7	12.246	9.5466	8.4513	7.8467	7.4604	7.1914	6.9928	6.8401	6.7188
8	11.259	8.6491	7.5910	7.0060	6.6318	6.3707	6.1776	6.0289	5.9106
9	10.561	8.0215	6.9919	6.4221	6.0569	5.8018	5.6129	5.4671	5.3511
10	10.044	7.5594	6.5523	5.9943	5.6363	5.3858	5.2001	5.0567	4.9424
11	9.6460	7.2057	6.2167	5.6683	5.3160	5.0692	4.8861	4.7445	4.6315
12	9.3302	6.9266	5.9526	5.4119	5.0643	4.8206	4.6395	4.4994	4.3875
13	9.0738	6.7010	5.7394	5.2053	4.8616	4.6204	4.4410	4.3021	4.1911
14	8.8616	6.5149	5.5639	5.0354	4.6950	4.4558	4.2779	4.1399	4.0297
15	8.6831	6.3589	5.4170	4.8932	4.5556	4.3183	4.1415	4.0045	3.8948
16	8.5310	6.2262	5.2922	4.7726	4.4374	4.2016	4.0259	3.8896	3.7804
17	8.3997	6.1121	5.1850	4.6690	4.3359	4.1015	3.9267	3.7910	3.6822
18	8.2854	6.0129	5.0919	4.5790	4.2479	4.0146	3.8406	3.7054	3.5971
19	8.1850	5.9259	5.0103	4.5003	4.1708	3.9386	3.7653	3.6305	3.5225
20	8.0960	5.8489	4.9382	4.4307	4.1027	3.8714	3.6987	3.5644	3.4567
21	8.0166	5.7804	4.8740	4.3688	4.0421	3.8117	3.6396	3.5056	3.3981
22	7.9454	5.7190	4.8166	4.3134	3.9880	3.7583	3.5867	3.4530	3.3458
23	7.8811	5.6637	4.7649	4.2635	3.9392	3.7102	3.5390	3.4057	3.2986
24	7.8229	5.6136	4.7181	4.2184	3.8951	3.6667	3.4959	3.3629	3.2560
25	7.7698	5.5680	4.6755	4.1774	3.8550	3.6272	3.4568	3.3239	3.2172
26	7.7213	5.5263	4.6366	4.1400	3.8183	3.5911	3.4210	3.2884	3.1818
27	7.6767	5.4881	4.6009	4.1056	3.7848	3.5580	3.3882	3.2558	3.1494
28	7.6356	5.4529	4.5681	4.0740	3.7539	3.5276	3.3581	3.2259	3.1195
29	7.5976	5.4205	4.5378	4.0449	3.7254	3.4995	3.3302	3.1982	3.0920
30	7.5625	5.3904	4.5097	4.0179	3.6990	3.4735	3.3045	3.1726	3.0665
40	7.3141	5.1785	4.3126	3.8283	3.5138	3.2910	3.1238	2.9930	2.8876
60	7.0771	4.9774	4.1259	3.6491	3.3389	3.1187	2.9530	2.8233	2.7185
120	6.8510	4.7865	3.9493	3.4796	3.1735	2.9559	2.7918	2.6629	2.5586
∞	6.6349	4.6052	3.7816	3.3192	3.0173	2.8020	2.6393	2.5113	2.4073

(*Continued*)

Table III (*Continued*)

					F Distribution: .01 Points					
ν_2 \ ν_1	10	12	15	20	24	30	40	60	120	∞
1	6055.8	6106.3	6157.3	6208.7	6234.6	6260.7	6286.8	6313.0	6339.4	6366.0
2	99.399	99.416	99.432	99.449	99.458	99.466	99.474	99.483	99.491	99.501
3	27.229	27.052	26.872	26.690	26.598	26.505	26.411	26.316	26.221	26.125
4	14.546	14.374	14.198	14.020	13.929	13.838	13.745	13.652	13.558	13.463
5	10.051	9.8883	9.7222	9.5527	9.4665	9.3793	9.2912	9.2020	9.1118	9.0204
6	7.8741	7.7183	7.5590	7.3958	7.3127	7.2285	7.1432	7.0568	6.9690	6.8801
7	6.6201	6.4691	6.3143	6.1554	6.0743	5.9921	5.9084	5.8236	5.7372	5.6495
8	5.8143	5.6668	5.5151	5.3591	5.2793	5.1981	5.1156	5.0316	4.9460	4.8588
9	5.2565	5.1114	4.9621	4.8080	4.7290	4.6486	4.5667	4.4831	4.3978	4.3105
10	4.8492	4.7059	4.5582	4.4054	4.3269	4.2469	4.1653	4.0819	3.9965	3.9090
11	4.5393	4.3974	4.2509	4.0990	4.0209	3.9411	3.8596	3.7761	3.6904	3.6025
12	4.2961	4.1553	4.0096	3.8584	3.7805	3.7008	3.6192	3.5355	3.4494	3.3608
13	4.1003	3.9603	3.8154	3.6646	3.5868	3.5070	3.4253	3.3413	3.2548	3.1654
14	3.9394	3.8001	3.6557	3.5052	3.4274	3.3476	3.2656	3.1813	3.0942	3.0040
15	3.8049	3.6662	3.5222	3.3719	3.2940	3.2141	3.1319	3.0471	2.9595	2.8684
16	3.6909	3.5527	3.4089	3.2588	3.1808	3.1007	3.0182	2.9330	2.8447	2.7528
17	3.5931	3.4552	3.3117	3.1615	3.0835	3.0032	2.9205	2.8348	2.7459	2.6530
18	3.5082	3.3706	3.2273	3.0771	2.9990	2.9185	2.8354	2.7493	2.6597	2.5660
19	3.4338	3.2965	3.1533	3.0031	2.9249	2.8422	2.7608	2.6742	2.5839	2.4893
20	3.3682	3.2311	3.0880	2.9377	2.8594	2.7785	2.6947	2.6077	2.5168	2.4212
21	3.3098	3.1729	3.0299	2.8796	2.8011	2.7200	2.6359	2.5484	2.4568	2.3603
22	3.2576	3.1209	2.9780	2.8274	2.7488	2.6675	2.5831	2.4951	2.4029	2.3055
23	3.2106	3.0740	2.9311	2.7805	2.7017	2.6202	2.5355	2.4471	2.3542	2.2559
24	3.1681	3.0316	2.8887	2.7380	2.6591	2.5773	2.4923	2.4035	2.3099	2.2107
25	3.1294	2.9931	2.8502	2.6993	2.6203	2.5383	2.4530	2.3637	2.2695	2.1694
26	3.0941	2.9579	2.8150	2.6640	2.5848	2.5026	2.4170	2.3273	2.2325	2.1315
27	3.0618	2.9256	2.7827	2.6316	2.5522	2.4699	2.3840	2.2938	2.1984	2.0965
28	3.0320	2.8959	2.7530	2.6017	2.5223	2.4397	2.3535	2.2629	2.1670	2.0642
29	3.0045	2.8685	2.7256	2.5742	2.4946	2.4118	2.3253	2.2344	2.1378	2.0342
30	2.9791	2.8431	2.7002	2.5487	2.4689	2.3860	2.2992	2.2079	2.1107	2.0062
40	2.8005	2.6648	2.5216	2.3689	2.2880	2.2034	2.1142	2.0194	1.9172	1.8047
60	2.6318	2.4961	2.3523	2.1978	2.1154	2.0285	1.9360	1.8363	1.7263	1.6006
120	2.4721	2.3363	2.1915	2.0346	1.9500	1.8600	1.7628	1.6557	1.5330	1.3805
∞	2.3209	2.1848	2.0385	1.8783	1.7908	1.6964	1.5923	1.4730	1.3246	1.0000

Table III *(Continued)*

	ν_1								
ν_2	1	2	3	4	5	6	7	8	9

F Distribution: .005 Points

ν_2	1	2	3	4	5	6	7	8	9
1	16211	20000	21615	22500	23056	23437	23715	23925	24091
2	198.50	199.00	199.17	199.25	199.30	199.33	199.36	199.37	199.39
3	55.552	49.799	47.467	46.195	45.392	44.838	44.434	44.126	43.882
4	31.333	26.284	24.259	23.155	22.456	21.975	21.622	21.352	21.139
5	22.785	18.314	16.530	15.556	14.940	14.513	14.200	13.961	13.772
6	18.635	14.544	12.917	12.028	11.464	11.073	10.786	10.566	10.391
7	16.236	12.404	10.882	10.050	9.5221	9.1554	8.8854	8.6781	8.5138
8	14.688	11.042	9.5965	8.8051	8.3018	7.9520	7.6942	7.4960	7.3386
9	13.614	10.107	8.7171	7.9559	7.4711	7.1338	6.8849	6.6933	6.5411
10	12.826	9.4270	8.0807	7.3428	6.8723	6.5446	6.3025	6.1159	5.9676
11	12.226	8.9122	7.6004	6.8809	6.4217	6.1015	5.8648	5.6821	5.5368
12	11.754	8.5096	7.2258	6.5211	6.0711	5.7570	5.5245	5.3451	5.2021
13	11.374	8.1865	6.9257	6.2335	5.7910	5.4819	5.2529	5.0761	4.9351
14	11.060	7.9217	6.6803	5.9984	5.5623	5.2574	5.0313	4.8566	4.7173
15	10.798	7.7008	6.4760	5.8029	5.3721	5.0708	4.8473	4.6743	4.5364
16	10.575	7.5138	6.3034	5.6378	5.2117	4.9134	4.6920	4.5207	4.3838
17	10.384	7.3536	6.1556	5.4967	5.0746	4.7789	4.5594	4.3893	4.2535
18	10.218	7.2148	6.0277	5.3746	4.9560	4.6627	4.4448	4.2759	4.1410
19	10.073	7.0935	5.9161	5.2681	4.8526	4.5614	4.3448	4.1770	4.0428
20	9.9439	6.9865	5.8177	5.1743	4.7616	4.4721	4.2569	4.0900	3.9564
21	9.8295	6.8914	5.7304	5.0911	4.6808	4.3931	4.1789	4.0128	3.8799
22	9.7271	6.8064	5.6524	5.0168	4.6088	4.3225	4.1094	3.9440	3.8116
23	9.6348	6.7300	5.5823	4.9500	4.5441	4.2591	4.0469	3.8822	3.7502
24	9.5513	6.6610	5.5190	4.8898	4.4857	4.2019	3.9905	3.8264	3.6949
25	9.4753	6.5982	5.4615	4.8351	4.4327	4.1500	3.9394	3.7758	3.6447
26	9.4059	6.5409	5.4091	4.7852	4.3844	4.1027	3.8928	3.7297	3.5989
27	9.3423	6.4885	5.3611	4.7396	4.3402	4.0594	3.8501	3.6875	3.5571
28	9.2838	6.4403	5.3170	4.6977	4.2996	4.0197	3.8110	3.6487	3.5186
29	9.2297	6.3958	5.2764	4.6591	4.2622	3.9830	3.7749	3.6130	3.4832
30	9.1797	6.3547	5.2388	4.6233	4.2276	3.9492	3.7416	3.5801	3.4505
40	8.8278	6.0664	4.9759	4.3738	3.9860	3.7129	3.5088	3.3498	3.2220
60	8.4946	5.7950	4.7290	4.1399	3.7600	3.4918	3.2911	3.1344	3.0083
120	8.1790	5.5393	4.4973	3.9207	3.5482	3.2849	3.0874	2.9330	2.8083
∞	7.8794	5.2983	4.2794	3.7151	3.3499	3.0913	2.8968	2.7444	2.6210

(Continued)

Table III *(Concluded)*

ν_2 \ ν_1	10	12	15	20	24	30	40	60	120	∞
1	24224	24426	24630	24836	24940	25044	25148	25253	25359	25465
2	199.40	199.42	199.43	199.45	199.46	199.47	199.47	199.48	199.49	199.51
3	43.686	43.387	43.085	42.778	42.622	42.466	42.308	42.149	41.989	41.829
4	20.967	20.705	20.438	20.167	20.030	19.892	19.752	19.611	19.468	19.325
5	13.618	13.384	13.146	12.903	12.780	12.656	12.530	12.402	12.274	12.144
6	10.250	10.034	9.8140	9.5888	9.4741	9.3583	9.2408	9.1219	9.0015	8.8793
7	8.3803	8.1764	7.9678	7.7540	7.6450	7.5345	7.4225	7.3088	7.1933	7.0760
8	7.2107	7.0149	6.8143	6.6082	6.5029	6.3961	6.2875	6.1772	6.0649	5.9505
9	6.4171	6.2274	6.0325	5.8318	5.7292	5.6248	5.5186	5.4104	5.3001	5.1875
10	5.8467	5.6613	5.4707	5.2740	5.1732	5.0705	4.9659	4.8592	4.7501	4.6385
11	5.4182	5.2363	5.0489	4.8552	4.7557	4.6543	4.5508	4.4450	4.3367	4.2256
12	5.0855	4.9063	4.7214	4.5299	4.4315	4.3309	4.2282	4.1229	4.0149	3.9039
13	4.8199	4.6429	4.4600	4.2703	4.1726	4.0727	3.9704	3.8655	3.7577	3.6465
14	4.6034	4.4281	4.2468	4.0585	3.9614	3.8619	3.7600	3.6553	3.5473	3.4359
15	4.4236	4.2498	4.0698	3.8826	3.7859	3.6867	3.5850	3.4803	3.3722	3.2602
16	4.2719	4.0994	3.9205	3.7342	3.6378	3.5388	3.4372	3.3324	3.2240	3.1115
17	4.1423	3.9709	3.7929	3.6073	3.5112	3.4124	3.3107	3.2058	3.0971	2.9839
18	4.0305	3.8599	3.6827	3.4977	3.4017	3.3030	3.2014	3.0962	2.9871	2.8732
19	3.9329	3.7631	3.5866	3.4020	3.3062	3.2075	3.1058	3.0004	2.8908	2.7762
20	3.8470	3.6779	3.5020	3.3178	3.2220	3.1234	3.0215	2.9159	2.8058	2.6904
21	3.7709	3.6024	3.4270	3.2431	3.1474	3.0488	2.9467	2.8408	2.7302	2.6140
22	3.7030	3.5350	3.3600	3.1764	3.0807	2.9821	2.8799	2.7736	2.6625	2.5455
23	3.6420	3.4745	3.2999	3.1165	3.0208	2.9221	2.8198	2.7132	2.6016	2.4837
24	3.5870	3.4199	3.2456	3.0624	2.9667	2.8679	2.7654	2.6585	2.5463	2.4276
25	3.5370	3.3704	3.1963	3.0133	2.9176	2.8187	2.7160	2.6088	2.4960	2.3765
26	3.4916	3.3252	3.1515	2.9685	2.8728	2.7738	2.6709	2.5633	2.4501	2.3297
27	3.4499	3.2839	3.1104	2.9275	2.8318	2.7327	2.6296	2.5217	2.4078	2.2867
28	3.4117	3.2460	3.0727	2.8899	2.7941	2.6949	2.5916	2.4834	2.3689	2.2469
29	3.3765	3.2111	3.0379	2.8551	2.7594	2.6601	2.5565	2.4479	2.3330	2.2102
30	3.3440	3.1787	3.0057	2.8230	2.7272	2.6278	2.5241	2.4151	2.2997	2.1760
40	3.1167	2.9531	2.7811	2.5984	2.5020	2.4015	2.2958	2.1838	2.0635	1.9318
60	2.9042	2.7419	2.5705	2.3872	2.2898	2.1874	2.0789	1.9622	1.8341	1.6885
120	2.7052	2.5439	2.3727	2.1881	2.0890	1.9839	1.8709	1.7469	1.6055	1.4311
∞	2.5188	2.3583	2.1868	1.9998	1.8983	1.7891	1.6691	1.5325	1.3637	1.0000

F Distribution: .005 Points

Source: Reproduced from Merrington, Maxine, and Thompson, Catherine M.: "Tables of Percentage Points of the Inverted Beta (*F*) Distribution," *Biometrika*, Vol. 33 (1943), by permission of *Biometrika* Trustees.

Table IV
Percentage points of the χ^2 distribution

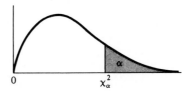

Probability of a Larger Value

df	.100	.050	.025	.010	.005
1	2.71	3.84	5.02	6.63	7.88
2	4.61	5.99	7.38	9.21	10.60
3	6.25	7.81	9.35	11.34	12.84
4	7.78	9.49	11.14	13.28	14.86
5	9.24	11.07	12.83	15.09	16.75
6	10.64	12.59	14.45	16.81	18.55
7	12.02	14.07	16.01	18.48	20.28
8	13.36	15.51	17.53	20.09	21.96
9	14.68	16.92	19.02	21.67	23.59
10	15.99	18.31	20.48	23.21	25.19
11	17.28	19.68	21.92	24.72	26.76
12	18.55	21.03	23.34	26.22	28.30
13	19.81	22.36	24.74	27.69	29.82
14	21.06	23.68	26.12	29.14	31.32
15	22.31	25.00	27.49	30.58	32.80
16	23.54	26.30	28.85	32.00	34.27
17	24.77	27.59	30.19	33.41	35.72
18	25.99	28.87	31.53	34.81	37.16
19	27.20	30.14	32.85	36.19	38.58
20	28.41	31.41	34.17	37.57	40.00
21	29.62	32.67	35.48	38.93	41.40
22	30.81	33.92	36.78	40.29	42.80
23	32.01	35.17	38.08	41.64	44.18
24	33.20	36.42	39.36	42.98	45.56
25	34.38	37.65	40.65	44.31	46.93
26	35.56	38.89	41.92	45.64	48.29
27	36.74	40.11	43.19	46.96	49.64
28	37.92	41.34	44.46	48.28	50.99
29	39.09	42.56	45.72	49.59	52.34
30	40.26	43.77	46.98	50.89	53.67
40	51.81	55.76	59.34	63.69	66.77
50	63.17	67.50	71.42	76.15	79.49
60	74.40	79.08	83.30	88.38	91.95
70	85.53	90.53	95.02	100.43	104.22
80	96.58	101.88	106.63	112.33	116.32
90	107.60	113.14	118.14	124.12	128.30
100	118.50	124.34	129.56	135.81	140.17

Source: Abridged from Thompson, Catherine M.: "Table of Percentage Points of the χ^2 Distribution," *Biometrika*, Vol. 32 (1942), p. 187, by permission by *Biometrika* Trustees.

Example: The probability is .05 of obtaining a χ^2 value in excess of 12.59 when the degrees of freedom are 6 (under the null hypothesis).

Table V
Durbin-Watson statistic (d). Significance points of d_L and d_U: 5%

n	$k' = 1$		$k' = 2$		$k' = 3$		$k' = 4$		$k' = 5$	
	d_L	d_U	d_L	d_U	d_L	d_U	d_L	d_U	d_L	d_U
15	1.08	1.36	0.95	1.54	0.82	1.75	0.69	1.97	0.56	2.21
16	1.10	1.37	0.98	1.54	0.86	1.73	0.74	1.93	0.62	2.15
17	1.13	1.38	1.02	1.54	0.90	1.71	0.78	1.90	0.67	2.10
18	1.16	1.39	1.05	1.53	0.93	1.69	0.82	1.87	0.71	2.06
19	1.18	1.40	1.08	1.53	0.97	1.68	0.86	1.85	0.75	2.02
20	1.20	1.41	1.10	1.54	1.00	1.68	0.90	1.83	0.79	1.99
21	1.22	1.42	1.13	1.54	1.03	1.67	0.93	1.81	0.83	1.96
22	1.24	1.43	1.15	1.54	1.05	1.66	0.96	1.80	0.86	1.94
23	1.26	1.44	1.17	1.54	1.08	1.66	0.99	1.79	0.90	1.92
24	1.27	1.45	1.19	1.55	1.10	1.66	1.01	1.78	0.93	1.90
25	1.29	1.45	1.21	1.55	1.12	1.66	1.04	1.77	0.95	1.89
26	1.30	1.46	1.22	1.55	1.14	1.65	1.06	1.76	0.98	1.88
27	1.32	1.47	1.24	1.56	1.16	1.65	1.08	1.76	1.01	1.86
28	1.33	1.48	1.26	1.56	1.18	1.65	1.10	1.75	1.03	1.85
29	1.34	1.48	1.27	1.56	1.20	1.65	1.12	1.74	1.05	1.84
30	1.35	1.49	1.28	1.57	1.21	1.65	1.14	1.74	1.07	1.83
31	1.36	1.50	1.30	1.57	1.23	1.65	1.16	1.74	1.09	1.83
32	1.37	1.50	1.31	1.57	1.24	1.65	1.18	1.73	1.11	1.82
33	1.38	1.51	1.32	1.58	1.26	1.65	1.19	1.73	1.13	1.81
34	1.39	1.51	1.33	1.58	1.27	1.65	1.21	1.73	1.15	1.81
35	1.40	1.52	1.34	1.58	1.28	1.65	1.22	1.73	1.16	1.80
36	1.41	1.52	1.35	1.59	1.29	1.65	1.24	1.73	1.18	1.80
37	1.42	1.53	1.36	1.59	1.31	1.66	1.25	1.72	1.19	1.80
38	1.43	1.54	1.37	1.59	1.32	1.66	1.26	1.72	1.21	1.79
39	1.43	1.54	1.38	1.60	1.33	1.66	1.27	1.72	1.22	1.79
40	1.44	1.54	1.39	1.60	1.34	1.66	1.29	1.72	1.23	1.79
45	1.48	1.57	1.43	1.62	1.38	1.67	1.34	1.72	1.29	1.78
50	1.50	1.59	1.46	1.63	1.42	1.67	1.38	1.72	1.34	1.77
55	1.53	1.60	1.49	1.64	1.45	1.68	1.41	1.72	1.38	1.77
60	1.55	1.62	1.51	1.65	1.48	1.69	1.44	1.73	1.41	1.77
65	1.57	1.63	1.54	1.66	1.50	1.70	1.47	1.73	1.44	1.77
70	1.58	1.64	1.55	1.67	1.52	1.70	1.49	1.74	1.46	1.77
75	1.60	1.65	1.57	1.68	1.54	1.71	1.51	1.74	1.49	1.77
80	1.61	1.66	1.59	1.69	1.56	1.72	1.53	1.74	1.51	1.77
85	1.62	1.67	1.60	1.70	1.57	1.72	1.55	1.75	1.52	1.77
90	1.63	1.68	1.61	1.70	1.59	1.73	1.57	1.75	1.54	1.78
95	1.64	1.69	1.62	1.71	1.60	1.73	1.58	1.75	1.56	1.78
100	1.65	1.69	1.63	1.72	1.61	1.74	1.59	1.76	1.57	1.78

n = number of observations.
k' = number of explanatory variables.

Table V   (*Continued*)
Durbin-Watson statistic (*d*). Significance points of d_L and d_U: 1%

n	$k' = 1$		$k' = 2$		$k' = 3$		$k' = 4$		$k' = 5$	
	d_L	d_U	d_L	d_U	d_L	d_U	d_L	d_U	d_L	d_L
15	0.81	1.07	0.70	1.25	0.59	1.46	0.49	1.70	0.39	1.96
16	0.84	1.09	0.74	1.25	0.63	1.44	0.53	1.66	0.44	1.90
17	0.87	1.10	0.77	1.25	0.67	1.43	0.57	1.63	0.48	1.85
18	0.90	1.12	0.80	1.26	0.71	1.42	0.61	1.60	0.52	1.80
19	0.93	1.13	0.83	1.26	0.74	1.41	0.65	1.58	0.56	1.77
20	0.95	1.15	0.86	1.27	0.77	1.41	0.68	1.57	0.60	1.74
21	0.97	1.16	0.89	1.27	0.80	1.41	0.72	1.55	0.63	1.71
22	1.00	1.17	0.91	1.28	0.83	1.40	0.75	1.54	0.66	1.69
23	1.02	1.19	0.94	1.29	0.86	1.40	0.77	1.53	0.70	1.67
24	1.04	1.20	0.96	1.30	0.88	1.41	0.80	1.53	0.72	1.66
25	1.05	1.21	0.98	1.30	0.90	1.41	0.83	1.52	0.75	1.65
26	1.07	1.22	1.00	1.31	0.93	1.41	0.85	1.52	0.78	1.64
27	1.09	1.23	1.02	1.32	0.95	1.41	0.88	1.51	0.81	1.63
28	1.10	1.24	1.04	1.32	0.97	1.41	0.90	1.51	0.83	1.62
29	1.12	1.25	1.05	1.33	0.99	1.42	0.92	1.51	0.85	1.61
30	1.13	1.26	1.07	1.34	1.01	1.42	0.94	1.51	0.88	1.61
31	1.15	1.27	1.08	1.34	1.02	1.42	0.96	1.51	0.90	1.60
32	1.16	1.28	1.10	1.35	1.04	1.43	0.98	1.51	0.92	1.60
33	1.17	1.29	1.11	1.36	1.05	1.43	1.00	1.51	0.94	1.59
34	1.18	1.30	1.13	1.36	1.07	1.43	1.01	1.51	0.95	1.59
35	1.19	1.31	1.14	1.37	1.08	1.44	1.03	1.51	0.97	1.59
36	1.21	1.32	1.15	1.38	1.10	1.44	1.04	1.51	0.99	1.59
37	1.22	1.32	1.16	1.38	1.11	1.45	1.06	1.51	1.00	1.59
38	1.23	1.33	1.18	1.39	1.12	1.45	1.07	1.52	1.02	1.58
39	1.24	1.34	1.19	1.39	1.14	1.45	1.09	1.52	1.03	1.58
40	1.25	1.34	1.20	1.40	1.15	1.46	1.10	1.52	1.05	1.58
45	1.29	1.38	1.24	1.42	1.20	1.48	1.16	1.53	1.11	1.58
50	1.32	1.40	1.28	1.45	1.24	1.49	1.20	1.54	1.16	1.59
55	1.36	1.43	1.32	1.47	1.28	1.51	1.25	1.55	1.21	1.59
60	1.38	1.45	1.35	1.48	1.32	1.52	1.28	1.56	1.25	1.60
65	1.41	1.47	1.38	1.50	1.35	1.53	1.31	1.57	1.28	1.61
70	1.43	1.49	1.40	1.52	1.37	1.55	1.34	1.58	1.31	1.61
75	1.45	1.50	1.42	1.53	1.39	1.56	1.37	1.59	1.34	1.62
80	1.47	1.52	1.44	1.54	1.42	1.57	1.39	1.60	1.36	1.62
85	1.48	1.53	1.46	1.55	1.43	1.58	1.41	1.60	1.39	1.63
90	1.50	1.54	1.47	1.56	1.45	1.59	1.43	1.61	1.41	1.64
95	1.51	1.55	1.49	1.57	1.47	1.60	1.45	1.62	1.42	1.64
100	1.52	1.56	1.50	1.58	1.48	1.60	1.46	1.63	1.44	1.65

n = number of observations.
k' = number of explanatory variables.

Source: Reproduced from *Biometrika*, Vol. 41 (1951), p. 173, by permission of *Biometrika* Trustees.

Table VI
Natural or Naperian logarithms

	0.000–0.499									
N	0	1	2	3	4	5	6	7	8	9
0.00	$-\infty$	6†	-6	-5	-5	-5	-5	-4	-4	-4
		.90776	.21461	.80914	.52146	.29832	.11600	.96185	.82831	.71053
.01	−4.60517	.50986	.42285	.34281	.26870	.19971	.13517	.07454	.01738	*.96332
.02	−3.91202	.86323	.81671	.77226	.72970	.68888	.64966	.61192	.57555	.54046
.03	.50656	.47377	.44202	.41125	.38139	.35241	.32424	.29684	.27017	.24419
.04	.21888	.19418	.17009	.14656	.12357	.10109	.07911	.05761	.03655	.01593
.05	−2.99573	.97593	.95651	.93746	.91877	.90042	.88240	.86470	.84731	.83022
.06	.81341	.79688	.78062	.76462	.74887	.73337	.71810	.70306	.68825	.67365
.07	.65926	.64508	.63109	.61730	.60369	.59027	.57702	.56395	.55105	.53831
.08	.52573	.51331	.50104	.48891	.47694	.46510	.45341	.44185	.43042	.41912
.09	.40795	.39690	.38597	.37516	.36446	.35388	.34341	.33304	.32279	.31264
0.10	−2.30259	.29263	.28278	.27303	.26336	.25379	.24432	.23493	.22562	.21641
.11	.20727	.19823	.18926	.18037	.17156	.16282	.15417	.14558	.13707	.12863
.12	.12026	.11196	.10373	.09557	.08747	.07944	.07147	.06357	.05573	.04794
.13	.04022	.03256	.02495	.01741	.00992	.00248	*.99510	*.98777	*.98050	*.97328
.14	−1.96611	.95900	.95193	.94491	.93794	.93102	.92415	.91732	.91054	.90381
.15	.89712	.89048	.88387	.87732	.87080	.86433	.85790	.85151	.84516	.83885
.16	.83258	.82635	.82016	.81401	.80789	.80181	.79577	.78976	.78379	.77786
.17	.77196	.76609	.76026	.75446	.74870	.74297	.73727	.73161	.72597	.72037
.18	.71480	.70926	.70375	.69827	.69282	.68740	.68201	.67665	.67131	.66601
.19	.66073	.65548	.65026	.64507	.63990	.63476	.62964	.62455	.61949	.61445
0.20	−1.60944	.60445	.59949	.59455	.58964	.58475	.57988	.57504	.57022	.56542
.21	.56065	.55590	.55117	.54646	.54178	.53712	.53248	.52786	.52326	.51868
.22	.51413	.50959	.50508	.50058	.49611	.49165	.48722	.48281	.47841	.47403
.23	.46968	.46534	.46102	.45672	.45243	.44817	.44392	.43970	.43548	.43129
.24	.42712	.42296	.41882	.41469	.41059	.40650	.40242	.39837	.39433	.39030
.25	.38629	.38230	.37833	.37437	.37042	.36649	.36258	.35868	.35480	.35093
.26	.34707	.34323	.33941	.33560	.33181	.32803	.32426	.32051	.31677	.31304
.27	.30933	.30564	.30195	.29828	.29463	.29098	.28735	.28374	.28013	.27654
.28	.27297	.26940	.26585	.26231	.25878	.25527	.25176	.24827	.24479	.24133
.29	.23787	.23443	.23100	.22758	.22418	.22078	.21740	.21402	.21066	.20731
0.30	−1.20397	.20065	.19733	.19402	.19073	.18744	.18417	.18091	.17766	.17441
.31	.17118	.16796	.16475	.16155	.15836	.15518	.15201	.14885	.14570	.14256
.32	.13943	.13631	.13320	.13010	.12701	.12393	.12086	.11780	.11474	.11170
.33	.10866	.10564	.10262	.09961	.09661	.09362	.09064	.08767	.08471	.08176
.34	.07881	.07587	.07294	.07002	.06711	.06421	.06132	.05843	.05555	.05268
.35	−1.04982	.04697	.04412	.04129	.03846	.03564	.03282	.03002	.02722	.02443
.36	.02165	.01888	.01611	.01335	.01060	.00786	.00512	.00239	*.99967	*.99696
.37	−0.99425	.99155	.98886	.98618	.98350	.98083	.97817	.97551	.97286	.97022
.38	.96758	.96496	.96233	.95972	.95711	.95451	.95192	.94933	.94675	.94418
.39	.94161	.93905	.93649	.93395	.93140	.92887	.92634	.92382	.92130	.91879
0.40	−0.91629	.91379	.91130	.90882	.90634	.90387	.90140	.89894	.89649	.89404
.41	.89160	.88916	.88673	.88431	.88189	.87948	.87707	.87467	.87227	.86988
.42	.86750	.86512	.86275	.86038	.85802	.85567	.85332	.85097	.84863	.84630
.43	.84397	.84165	.83933	.83702	.83471	.83241	.83011	.82782	.82554	.82326
.44	.82098	.81871	.81645	.81419	.81193	.80968	.80744	.80520	.80296	.80073
.45	.79851	.79629	.79407	.79186	.78966	.78746	.78526	.78307	.78089	.77871
.46	.77653	.77436	.77219	.77003	.76787	.76572	.76357	.76143	.75929	.75715
.47	.75502	.75290	.75078	.74866	.74655	.74444	.74234	.74024	.73814	.73605
.48	.73397	.73189	.72981	.72774	.72567	.72361	.72155	.71949	.71744	.71539
.49	.71335	.71131	.70928	.70725	.70522	.70320	.70118	.69917	.69716	.69515

† Note that the characteristics are given *above* the mantissa for the first line. In the second and following lines they are given at the left.

Table VI   (*Continued*)

N	0	1	2	3	4	5	6	7	8	9
					0.500–0.999					
0.50	−0.69315	.69115	.68916	.68717	.68518	.68320	.68122	.67924	.67727	.67531
.51	.67334	.67139	.66943	.66748	.66553	.66359	.66165	.65971	.65778	.65585
.52	.65393	.65201	.65009	.64817	.64626	.64436	.64245	.64055	.63866	.63677
.53	.63488	.63299	.63111	.62923	.62736	.62549	.62362	.62176	.61990	.61804
.54	.61619	.61434	.61249	.61065	.60881	.60697	.60514	.60331	.60148	.59966
.55	.59784	.59602	.59421	.59240	.59059	.58879	.58699	.58519	.58340	.58161
.56	.57982	.57803	.57625	.57448	.57270	.57093	.56916	.56740	.56563	.56387
.57	.56212	.56037	.55862	.55687	.55513	.55339	.55165	.54991	.54818	.54645
.58	.54473	.54300	.54128	.53957	.53785	.53614	.53444	.53273	.53103	.52933
.59	.52763	.52594	.52425	.52256	.52088	.51919	.51751	.51584	.51416	.51249
0.60	−0.51083	.50916	.50750	.50584	.50418	.50253	.50088	.49923	.49758	.49594
.61	.49430	.49266	.49102	.48939	.48776	.48613	.48451	.48289	.48127	.47965
.62	.47804	.47642	.47482	.47321	.47160	.47000	.46840	.46681	.46522	.46362
.63	.46204	.46045	.45887	.45728	.45571	.45413	.45256	.45099	.44942	.44785
.64	.44629	.44473	.44317	.44161	.44006	.43850	.43696	.43541	.43386	.43232
.65	.43078	.42925	.42771	.42618	.42465	.42312	.42159	.42007	.41855	.41703
.66	.41552	.41400	.41249	.41098	.40947	.40797	.40647	.40497	.40347	.40197
.67	.40048	.39899	.39750	.39601	.39453	.39304	.39156	.39008	.38861	.38713
.68	.38566	.38419	.38273	.38126	.37980	.37834	.37688	.37542	.37397	.37251
.69	.37106	.36962	.36817	.36673	.36528	.36384	.36241	.36097	.35954	.35810
0.70	−0.35667	.35525	.35382	.35240	.35098	.34956	.34814	.34672	.34531	.34390
.71	.34249	.34108	.33968	.33827	.33687	.33547	.33408	.33268	.33129	.32989
.72	.32850	.32712	.32573	.32435	.32296	.32158	.32021	.31883	.31745	.31608
.73	.31471	.31334	.31197	.31061	.30925	.30788	.30653	.30517	.30381	.30246
.74	.30111	.29975	.29841	.29706	.29571	.29437	.29303	.29169	.29035	.28902
.75	.28768	.28635	.28502	.28369	.28236	.28104	.27971	.27839	.27707	.27575
.76	.27444	.27312	.27181	.27050	.26919	.26788	.26657	.26527	.26397	.26266
.77	.26136	.26007	.25877	.25748	.25618	.25489	.25360	.25231	.25103	.24974
.78	.24846	.24718	.24590	.24462	.24335	.24207	.24080	.23953	.23826	.23699
.79	.23572	.23446	.23319	.23193	.23067	.22941	.22816	.22690	.22565	.22439
0.80	−0.22314	.22189	.22065	.21940	.21816	.21691	.21567	.21443	.21319	.21196
.81	.21072	.20949	.20825	.20702	.20579	.20457	.20334	.20212	.20089	.19967
.82	.19845	.19723	.19601	.19480	.19358	.19237	.19116	.18995	.18874	.18754
.83	.18633	.18513	.18392	.18272	.18152	.18032	.17913	.17793	.17674	.17554
.84	.17435	.17316	.17198	.17079	.16960	.16842	.16724	.16605	.16487	.16370
.85	−0.16252	.16134	.16017	.15900	.15782	.15665	.15548	.15432	.15315	.15199
.86	.15082	.14966	.14850	.14734	.14618	.14503	.14387	.14272	.14156	.14041
.87	.13926	.13811	.13697	.13582	.13467	.13353	.13239	.13125	.13011	.12897
.88	.12783	.12670	.12556	.12443	.12330	.12217	.12104	.11991	.11878	.11766
.89	.11653	.11541	.11429	.11317	.11205	.11093	.10981	.10870	.10759	.10647
0.90	−0.10536	.10425	.10314	.10203	.10093	.09982	.09872	.09761	.09651	.09541
.91	.09431	.09321	.09212	.09102	.08992	.08883	.08774	.08665	.08556	.08447
.92	.08338	.08230	.08121	.08013	.07904	.07796	.07688	.07580	.07472	.07365
.93	.07257	.07150	.07042	.06935	.06828	.06721	.06614	.06507	.06401	.06294
.94	.06188	.06081	.05975	.05869	.05763	.05657	.05551	.05446	.05340	.05235
.95	.05129	.05024	.04919	.04814	.04709	.04604	.04500	.04395	.04291	.04186
.96	.04082	.03978	.03874	.03770	.03666	.03563	.03459	.03356	.03252	.03149
.97	.03046	.02943	.02840	.02737	.02634	.02532	.02429	.02327	.02225	.02122
.98	.02020	.01918	.01816	.01715	.01613	.01511	.01410	.01309	.01207	.01106
.99	.01005	.00904	.00803	.00702	.00602	.00501	.00401	.00300	.00200	.00100

(*Continued*)

Table VI (*Continued*)

To find the natural logarithm of a number which is $\frac{1}{10}, \frac{1}{100}, \frac{1}{1000}$, etc. of a number whose logarithm is given, subtract from the given logarithm $\log_e 10$, $2 \log_e 10$, $3 \log_e 10$, etc.

To find the natural logarithm of a number which is 10, 100, 1000, etc. times a number whose logarithm is given, add to the given logarithm $\log_e 10$, $2 \log_e 10$, $3 \log_e 10$, etc.

$\log_e 10 =$ 2.30258 50930	$6 \log_e 10 =$ 13.81551 05580
$2 \log_e 10 =$ 4.60517 01860	$7 \log_e 10 =$ 16.11809 56510
$3 \log_e 10 =$ 6.90775 52790	$8 \log_e 10 =$ 18.42068 07440
$4 \log_e 10 =$ 9.21034 03720	$9 \log_e 10 =$ 20.72326 58369
$5 \log_e 10 =$ 11.51292 54650	$10 \log_e 10 =$ 23.02585 09299

See preceding table for logarithms for numbers between 0.000 and 0.999.

1.00–4.99

N	0	1	2	3	4	5	6	7	8	9
1.0	0.00000	.00995	.01980	.02956	.03922	.04879	.05827	.06766	.07696	.08618
.1	.09531	.10436	.11333	.12222	.13103	.13976	.14842	.15700	.16551	.17395
.2	.18232	.19062	.19885	.20701	.21511	.22314	.23111	.23902	.24686	.25464
.3	.26236	.27003	.27763	.28518	.29267	.30010	.30748	.31481	.32208	.32930
.4	.33647	.34359	.35066	.35767	.36464	.37156	.37844	.38526	.39204	.39878
.5	.40547	.41211	.41871	.42527	.43178	.43825	.44469	.45108	.45742	.46373
.6	.47000	.47623	.48243	.48858	.49470	.50078	.50682	.51282	.51879	.52473
.7	.53063	.53649	.54232	.54812	.55389	.55962	.56531	.57098	.57661	.58222
.8	.58779	.59333	.59884	.60432	.60977	.61519	.62058	.62594	.63127	.63658
.9	.64185	.64710	.65233	.65752	.66269	.66783	.67294	.67803	.68310	.68813
2.0	0.69315	.69813	.70310	.70804	.71295	.71784	.72271	.72755	.73237	.73716
.1	.74194	.74669	.75142	.75612	.76081	.76547	.77011	.77473	.77932	.78390
.2	.78846	.79299	.79751	.80200	.80648	.81093	.81536	.81978	.82418	.82855
.3	.83291	.83725	.84157	.84587	.85015	.85442	.85866	.86280	.86710	.87129
.4	.87547	.87963	.88377	.88789	.89200	.89609	.90016	.90422	.90826	.91228
.5	.91629	.92028	.92426	.92822	.93216	.93609	.94001	.94391	.94779	.95166
.6	.95551	.95935	.96317	.96698	.97078	.97456	.97833	.98208	.98582	.98954
.7	.99325	.99695	*.00063	*.00430	*.00796	*.01160	*.01523	*.01885	*.02245	*.02604
.8	1.02962	.03318	.03674	.04028	.04380	.04732	.05082	.05431	.05779	.06126
.9	.06471	.06815	.07158	.07500	.07841	.08181	.08519	.08856	.09192	.09527
3.0	1.09861	.10194	.10526	.10856	.11186	.11514	.11841	.12168	.12493	.12817
.1	.13140	.13462	.13783	.14103	.14422	.14740	.15057	.15373	.15688	.16002
.2	.16315	.16627	.16938	.17248	.17557	.17865	.18173	.18479	.18784	.19089
.3	.19392	.19695	.19996	.20297	.20597	.20896	.21194	.21491	.21788	.22083
.4	.22378	.22671	.22964	.23256	.23647	.23837	.24127	.24415	.24703	.24990
.5	.25276	.25562	.25846	.26130	.26413	.26695	.26976	.27257	.27536	.27815
.6	.28093	.28371	.28647	.28923	.29198	.29473	.29746	.30019	.30291	.30563
.7	.30833	.31103	.31372	.31641	.31909	.32176	.32442	.32708	.32972	.33237
.8	.33500	.33763	.34025	.34286	.34547	.34807	.35067	.35325	.35584	.35841
.9	.36098	.36354	.36609	.36864	.37118	.37372	.37624	.37877	.38128	.38379
4.0	1.38629	.38879	.39128	.39377	.39624	.39872	.40118	.40364	.40610	.40854
.1	.41099	.41342	.41585	.41828	.42070	.42311	.42552	.42792	.43031	.43270
.2	.43508	.43746	.43984	.44220	.44456	.44692	.44927	.45161	.45395	.45629
.3	.45862	.46094	.46326	.46557	.46787	.47018	.47247	.47476	.47705	.47933
.4	.48160	.48387	.48614	.48840	.49065	.49290	.49515	.49739	.49962	.50185
.5	.50408	.50630	.50851	.51072	.51293	.51513	.51732	.51951	.52170	.52388
.6	.52606	.52823	.53039	.53256	.53471	.53687	.53902	.54116	.54330	.54543
.7	.54756	.54969	.55181	.55393	.55604	.55814	.56025	.56235	.56444	.56653
.8	.56862	.57070	.57277	.57485	.57691	.57898	.58104	.58309	.58515	.58719
.9	.58924	.59127	.59331	.59534	.59737	.59939	.60141	.60342	.60543	.60744

Table VI (*Continued*)

N	0	1	2	3	4	5	6	7	8	9
					5.00–9.99					
5.0	1.60944	.61144	.61343	.61542	.61741	.61939	.62137	.62334	.62531	.62728
.1	.62924	.63120	.63315	.63511	.63705	.63900	.64094	.64287	.64481	.64673
.2	.64866	.65058	.65250	.65441	.65632	.65823	.66013	.66203	.66393	.66582
.3	.66771	.66959	.67147	.67335	.67523	.67710	.67896	.68083	.68269	.68455
.4	.68640	.68825	.69010	.69194	.69378	.69562	.69745	.69928	.70111	.70293
.5	.70475	.70656	.70838	.71019	.71199	.71380	.71560	.71740	.71919	.72098
.6	.72277	.72455	.72633	.72811	.72988	.73166	.73342	.73519	.73695	.73871
.7	.74047	.74222	.74397	.74572	.74746	.74920	.75094	.75267	.75440	.75613
.8	.75786	.75958	.76130	.76302	.76473	.76644	.76815	.76985	.77156	.77326
.9	.77495	.77665	.77834	.78002	.78171	.78339	.78507	.78675	.78842	.79009
6.0	1.79176	.79342	.79509	.79675	.79840	.80006	.80171	.80336	.80500	.80665
.1	.80829	.80993	.81156	.81319	.81482	.81645	.81808	.81970	.82132	.82294
.2	.82455	.82616	.82777	.82938	.83098	.83258	.83418	.83578	.83737	.83896
.3	.84055	.84214	.84372	.84530	.84688	.84845	.85003	.85160	.85317	.85473
.4	.85630	.85786	.85942	.86097	.86253	.86408	.86563	.86718	.86872	.87026
.5	.87180	.87334	.87487	.87641	.87794	.87947	.88099	.88251	.88403	.88555
.6	.88707	.88858	.89010	.89160	.89311	.89462	.89612	.89762	.89912	.90061
.7	.90211	.90360	.90509	.90658	.90806	.90954	.91102	.91250	.91398	.91545
.8	.91692	.91839	.91986	.92132	.92279	.92425	.92571	.92716	.92862	.93007
.9	.93152	.93297	.93442	.93586	.93730	.93874	.94018	.94162	.94305	.94448
7.0	1.94591	.94734	.94876	.95019	.95161	.95303	.95445	.95586	.95727	.95869
.1	.96009	.96150	.96291	.96431	.96571	.96711	.96851	.96991	.97130	.97269
.2	.97408	.97547	.97685	.97824	.97962	.98100	.98238	.98376	.98513	.98650
.3	.98787	.98924	.99061	.99198	.99334	.99470	.99606	.99742	.99877	*.00013
.4	2.00148	.00283	.00418	.00553	.00687	.00821	.00956	.01089	.01223	.01357
.5	.01490	.01624	.01757	.01890	.02022	.02155	.02287	.02419	.02551	.02683
.6	.02815	.02946	.03078	.03209	.03340	.03471	.03601	.03732	.03862	.03992
.7	.04122	.04252	.04381	.04511	.04640	.04769	.04898	.05027	.05156	.05284
.8	.05412	.05540	.05668	.05796	.05924	.06051	.06179	.06306	.06433	.06560
.9	.06686	.06813	.06939	.07065	.07191	.07317	.07443	.07568	.07694	.07819
8.0	2.07944	.08069	.08194	.08318	.08443	.08567	.08691	.08815	.08939	.09063
.1	.09186	.09310	.09433	.09556	.09679	.09802	.09924	.10047	.10169	.10291
.2	.10413	.10535	.10657	.10779	.10900	.11021	.11142	.11263	.11384	.11505
.3	.11626	.11746	.11866	.11986	.12106	.12226	.12346	.12465	.12585	.12704
.4	.12823	.12942	.13061	.13180	.13298	.13417	.13535	.13653	.13771	.13889
.5	.14007	.14124	.14242	.14359	.14476	.14593	.14710	.14827	.14943	.15060
.6	.15176	.15292	.15409	.15524	.15640	.15756	.15871	.15987	.16102	.16217
.7	.16332	.16447	.16562	.16677	.16791	.16905	.17020	.17134	.17248	.17361
.8	.17475	.17589	.17702	.17816	.17929	.18042	.18155	.18267	.18380	.18493
.9	.18605	.18717	.18830	.18942	.19054	.19165	.19277	.19389	.19500	.19611
9.0	2.19722	.19834	.19944	.20055	.20166	.20276	.20387	.20497	.20607	.20717
.1	.20827	.20937	.21047	.21157	.21266	.21375	.21485	.21594	.21703	.21812
.2	.21920	.22029	.22138	.22246	.22354	.22462	.22570	.22678	.22786	.22894
.3	.23001	.23109	.23216	.23324	.23431	.23538	.23645	.23751	.23858	.23965
.4	.24071	.24177	.24284	.24390	.24496	.24601	.24707	.24813	.24918	.25024
.5	.25129	.25234	.25339	.25444	.25549	.25654	.25759	.25863	.25968	.26072
.6	.26176	.26280	.26384	.26488	.26592	.26696	.26799	.26903	.27006	.27109
.7	.27213	.27316	.27419	.27521	.27624	.27727	.27829	.27932	.28034	.28136
.8	.28238	.28340	.28442	.28544	.28646	.28747	.28849	.28950	.29051	.29152
.9	.29253	.29354	.29455	.29556	.29657	.29757	.29858	.29958	.30058	.30158

(*Continued*)

Table VI (*Continued*)

Constants

$$\log_e 10 = 2.30258\ 50930 \qquad 6\log_e 10 = 13.81551\ 05580$$
$$2\log_e 10 = 4.60517\ 01860 \qquad 7\log_e 10 = 16.11809\ 56510$$
$$3\log_e 10 = 6.90775\ 52790 \qquad 8\log_e 10 = 18.42068\ 07440$$
$$4\log_e 10 = 9.21034\ 03720 \qquad 9\log_e 10 = 20.72326\ 58369$$
$$5\log_e 10 = 11.51292\ 54650 \qquad 10\log_e 10 = 23.02585\ 09299$$

10.0–49.9

N	0	1	2	3	4	5	6	7	8	9
10.	2.30259	.31254	.32239	.33214	.34181	.35138	.36085	.37024	.37955	.38876
11.	.39790	.40695	.41591	.42480	.43361	.44235	.45101	.45959	.46810	.47654
12.	.48491	.49321	.50144	.50960	.51770	.52573	.53370	.54160	.54945	.55723
13.	.56495	.57261	.58022	.58776	.59525	.60269	.61007	.61740	.62467	.63189
14.	.63906	.64617	.65324	.66026	.66723	.67415	.68102	.68785	.69463	.70136
15.	.70805	.71469	.72130	.72785	.73437	.74084	.74727	.75366	.76001	.76632
16.	.77259	.77882	.78501	.79117	.79728	.80336	.80940	.81541	.82138	.82731
17.	.83321	.83908	.84491	.85071	.85647	.86220	.86790	.87356	.87920	.88480
18.	.89037	.89591	.90142	.90690	.91235	.91777	.92316	.92852	.93386	.93916
19.	.94444	.94969	.95491	.96011	.96527	.97041	.97553	.98062	.98568	.99072
20.	2.99573	*.00072	*.00568	*.01062	*.01553	*.02042	*.02529	*.03013	*.03495	*.03975
21.	3.04452	.04927	.05400	.05871	.06339	.06805	.07269	.07731	.08191	.08649
22.	.09104	.09558	.10009	.10459	.10906	.11352	.11795	.12236	.12676	.13114
23.	.13549	.13983	.14415	.14845	.15274	.15700	.16125	.16548	.16969	.17388
24.	.17805	.18221	.18635	.19048	.19458	.19867	.20275	.20680	.21084	.21487
25.	.21888	.22287	.22684	.23080	.23475	.23868	.24259	.24649	.25037	.25424
26.	.25810	.26194	.26576	.26957	.27336	.27714	.28091	.28466	.28840	.29213
27.	.29584	.29953	.30322	.30689	.31054	.31419	.31782	.32143	.32504	.32863
28.	.33220	.33577	.33932	.34286	.34639	.34990	.35341	.35690	.36038	.36384
29.	.36730	.37074	.37417	.37759	.38099	.38439	.38777	.39115	.39451	.39786
30.	3.40120	.40453	.40784	.41115	.41444	.41773	.42100	.42426	.42751	.43076
31.	.43399	.43721	.44042	.44362	.44681	.44999	.45316	.45632	.45947	.46261
32.	.46574	.46886	.47197	.47507	.47816	.48124	.48431	.48738	.49043	.49347
33.	.49651	.49953	.50255	.50556	.50856	.51155	.51453	.51750	.52046	.52342
34.	.52636	.52930	.53223	.53515	.53806	.54096	.54385	.54674	.54962	.55249
35.	.55535	.55820	.56105	.56388	.56671	.56953	.57235	.57515	.57795	.58074
36.	.58352	.58629	.58906	.59182	.59457	.59731	.60005	.60278	.60550	.60821
37.	.61092	.61362	.61631	.61899	.62167	.62434	.62700	.62966	.63231	.63495
38.	.63759	.64021	.64284	.64545	.64806	.65066	.65325	.65584	.65842	.66099
39.	.66356	.66612	.66868	.67122	.67377	.67630	.67883	.68135	.68387	.68638
40.	3.68888	.69138	.69387	.69635	.69883	.70130	.70377	.70623	.70868	.71113
41.	.71357	.71601	.71844	.72086	.72328	.72569	.72810	.73050	.73290	.73529
42.	.73767	.74005	.74242	.74479	.74715	.74950	.75185	.75420	.75654	.75887
43.	.76120	.76352	.76584	.76815	.77046	.77276	.77506	.77735	.77963	.78191
44.	.78419	.78646	.78872	.79098	.79324	.79549	.79773	.79997	.80221	.80444
45.	.80666	.80888	.81110	.81331	.81551	.81771	.81991	.82210	.82428	.82647
46.	.82864	.83081	.83298	.83514	.83730	.83945	.84160	.84374	.84588	.84802
47.	.85015	.85227	.85439	.85651	.85862	.86073	.86283	.86493	.86703	.86912
48.	.87120	.87328	.87536	.87743	.87950	.88156	.88362	.88568	.88773	.88978
49.	.89182	.89386	.89589	.89792	.89995	.90197	.90399	.90600	.90801	.91002

Table VI *(Continued)*

						50.0–99.9				
N	0	1	2	3	4	5	6	7	8	9
50.	3.91202	.91402	.91602	.91801	.91999	.92197	.92395	.92593	.92790	.92986
51.	.93183	.93378	.93574	.93769	.93964	.94158	.94352	.94546	.94739	.94932
52.	.95124	.95316	.95508	.95700	.95891	.96081	.96272	.96462	.96651	.96840
53.	.97029	.97218	.97406	.97594	.97781	.97968	.98155	.98341	.98527	.98713
54.	.98898	.99083	.99268	.99452	.99636	.99820	*.00003	*.00186	*.00369	*.00551
55.	4.00733	.00915	.01096	.01277	.01458	.01638	.01818	.01998	.02177	.02356
56.	.02535	.02714	.02892	.03069	.03247	.03424	.03601	.03777	.03954	.04130
57.	.04305	.04480	.04655	.04830	.05004	.05178	.05352	.05526	.05699	.05872
58.	.06044	.06217	.06389	.06560	.06732	.06903	.07073	.07244	.07414	.07584
59.	.07754	.07923	.08092	.08261	.08429	.08598	.08766	.08933	.09101	.09268
60.	4.09434	.09601	.09767	.09933	.10099	.10264	.10429	.10594	.10759	.10923
61.	.11087	.11251	.11415	.11578	.11741	.11904	.12066	.12228	.12390	.12552
62.	.12713	.12875	.13036	.13196	.13357	.13517	.13677	.13836	.13996	.14155
63.	.14313	.14472	.14630	.14789	.14946	.15104	.15261	.15418	.15575	.15732
64.	.15888	.16044	.16200	.16356	.16511	.16667	.16821	.16976	.17131	.17285
65.	.17439	.17592	.17746	.17899	.18052	.18205	.18358	.18510	.18662	.18814
66.	.18965	.19117	.19268	.19419	.19570	.19720	.19870	.20020	.20170	.20320
67.	.20469	.20618	.20767	.20916	.21065	.21213	.21361	.21509	.21656	.21804
68.	.21951	.22098	.22244	.22391	.22537	.22683	.22829	.22975	.23120	.23266
69.	.23411	.23555	.23700	.23844	.23989	.24133	.24276	.24420	.24563	.24707
70.	4.24850	.24992	.25135	.25277	.25419	.25561	.25703	.25845	.25986	.26127
71.	.26268	.26409	.26549	.26690	.26830	.26970	.27110	.27249	.27388	.27528
72.	.27667	.27805	.27944	.28082	.28221	.28359	.28496	.28634	.28772	.28909
73.	.29046	.29183	.29320	.29456	.29592	.29729	.29865	.30000	.30136	.30271
74.	.30407	.30542	.30676	.30811	.30946	.31080	.31214	.31348	.31482	.31615
75.	.31749	.31882	.32015	.32149	.32281	.32413	.32546	.32678	.32810	.32942
76.	.33073	.33205	.33336	.33467	.33598	.33729	.33860	.33990	.34120	.34251
77.	.34381	.34510	.34640	.34769	.34899	.35028	.35157	.35286	.35414	.35543
78.	.35671	.35800	.35927	.36055	.36182	.36310	.36437	.36564	.36691	.36818
79.	.36945	.37071	.37198	.37324	.37450	.37576	.37701	.37827	.37952	.38078
80.	4.38203	.38328	.38452	.38577	.38701	.38826	.38950	.39074	.39198	.39321
81.	.39445	.39568	.29692	.39815	.39938	.40060	.40183	.40305	.40428	.40550
82.	.40672	.40794	.40916	.41037	.41159	.41280	.41401	.41522	.41643	.41764
83.	.41884	.42004	.42125	.42245	.42365	.42485	.42604	.42724	.42843	.42963
84.	.43082	.43201	.43319	.43438	.43557	.43675	.43793	.43912	.44030	.44147
85.	.44265	.44383	.44500	.44617	.44735	.44852	.44969	.45085	.45202	.45318
86.	.45435	.45551	.45667	.45783	.45899	.46014	.46130	.46245	.46361	.46476
87.	.46591	.46706	.46820	.46935	.47050	.47164	.47278	.47392	.47506	.47620
88.	.47734	.47847	.47961	.48074	.48187	.48300	.48413	.48526	.48639	.48751
89.	.48864	.48976	.49088	.49200	.49312	.49424	.49536	.49647	.49758	.49870
90.	4.49981	.50092	.50203	.50314	.50424	.50535	.50645	.50756	.50866	.50976
91.	.51086	.51196	.51305	.51415	.51525	.51634	.51743	.51852	.51961	.52070
92.	.52179	.52287	.52396	.52504	.52613	.52721	.52829	.52937	.53045	.53152
93.	.53260	.53367	.53475	.53582	.53689	.53796	.53903	.54010	.54116	.54223
94.	.54329	.54436	.54542	.54648	.54754	.54860	.54966	.55071	.55177	.55282
95.	.55388	.55493	.55598	.55703	.55808	.55913	.56017	.56122	.56226	.56331
96.	.56435	.56539	.56643	.56747	.56851	.56954	.57058	.57161	.57265	.57368
97.	.57471	.57574	.57677	.57780	.57883	.57985	.58088	.58190	.58292	.58395
98.	.58497	.58599	.58701	.58802	.58904	.59006	.59107	.59208	.59310	.59411
99.	.59512	.59613	.59714	.59815	.59915	.60016	.60116	.60217	.60317	.60417

(Continued)

Table VI   (*Continued*)

N	0	1	2	3	4	5	6	7	8	9
					0–499					
0	∞	0.00000	0.69315	1.09861	.38629	.60944	.79176	.94591	*.07944	*.19722
1	2.30259	.39790	.48491	.56495	.63906	.70805	.77259	.83321	.89037	.94444
2	.99573	*.04452	*.09104	*.13549	*.17805	*.21888	*.25810	*.29584	*.33220	*.36730
3	3.40120	.43399	.46574	.49651	.52636	.55535	.58352	.61092	.63759	.66356
4	.68888	.71357	.73767	.76120	.78419	.80666	.82864	.85015	.87120	.89182
5	.91202	.93183	.95124	.97029	.98898	*.00733	*.02535	*.04305	*.06044	*.07754
6	4.09434	.11087	.12713	.14313	.15888	.17439	.18965	.20469	.21951	.23411
7	.24850	.26268	.27667	.29046	.30407	.31749	.33073	.34381	.35671	.36945
8	.38203	.39445	.40672	.41884	.43082	.44265	.45435	.46591	.47734	.48864
9	.49981	.51086	.52179	.53260	.54329	.55388	.56435	.57471	.58497	.59512
10	4.60517	.61512	.62497	.63473	.64439	.65396	.66344	.67283	.68213	.69135
11	.70048	.70953	.71850	.72739	.73620	.74493	.75359	.76217	.77068	.77912
12	.78749	.79579	.80402	.81218	.82028	.82831	.83628	.84419	.85203	.85981
13	.86753	.87520	.88280	.89035	.89784	.90527	.91265	.91998	.92725	.93447
14	.94164	.94876	.95583	.96284	.96981	.97673	.98361	.99043	.99721	*.00395
15	5.01064	.01728	.02388	.03044	.03695	.04343	.04986	.05625	.06260	.06890
16	.07517	.08140	.08760	.09375	.09987	.10595	.11199	.11799	.12396	.12990
17	.13580	.14166	.14749	.15329	.15906	.16479	.17048	.17615	.18178	.18739
18	.19296	.19850	.20401	.20949	.21494	.22036	.22575	.23111	.23644	.24175
19	.24702	.25227	.25750	.26269	.26786	.27300	.27811	.28320	.28827	.29330
20	5.29832	.30330	.30827	.31321	.31812	.32301	.32788	.33272	.33754	.34233
21	.34711	.35186	.35659	.36129	.36598	.37064	.37528	.37990	.38450	.38907
22	.39363	.39816	.40268	.40717	.41165	.41610	.42053	.42495	.42935	.43372
23	.43808	.44242	.44674	.45104	.45532	.45959	.46383	.46806	.47227	.47646
24	.48064	.48480	.48894	.49306	.49717	.50126	.50533	.50939	.51343	.51745
25	.52146	.52545	.52943	.53339	.53733	.54126	.54518	.54908	.55296	.55683
26	.56068	.56452	.56834	.57215	.57595	.57973	.58350	.58725	.59099	.59471
27	.59842	.60212	.60580	.60947	.61313	.61677	.62040	.62402	.62762	.63121
28	.63479	.63835	.64191	.64545	.64897	.65249	.65599	.65948	.66296	.66643
29	.66988	.67332	.67675	.68017	.68358	.68698	.69036	.69373	.69709	.70044
30	5.70378	.70711	.71043	.71373	.71703	.72031	.72359	.72685	.73010	.73334
31	.73657	.73979	.74300	.74620	.74939	.75257	.75574	.75890	.76205	.76519
32	.76832	.77144	.77455	.77765	.78074	.78383	.78690	.78996	.79301	.79606
33	.79909	.80212	.80513	.80814	.81114	.81413	.81711	.82008	.82305	.82600
34	.82895	.83188	.83481	.83773	.84064	.84354	.84644	.84932	.85220	.85507
35	.85793	.86079	.86363	.86647	.86930	.87212	.87493	.87774	.88053	.88332
36	.88610	.88888	.89164	.89440	.89715	.89990	.90263	.90536	.90808	.91080
37	.91350	.91620	.91889	.92158	.92426	.92693	.92959	.93225	.93489	.93754
38	.94017	.94280	.94542	.94803	.95064	.95324	.95584	.95842	.96101	.96358
39	.96615	.96871	.97126	.97381	.97635	.97889	.98141	.98394	.98645	.98896
40	5.99146	.99396	.99645	.99894	*.00141	*.00389	*.00635	*.00881	*.01127	*.01372
41	6.01616	.01859	.02102	.02345	.02587	.02828	.03069	.03309	.03548	.03787
42	.04025	.04263	.04501	.04737	.04973	.05209	.05444	.05678	.05912	.06146
43	.06379	.06611	.06843	.07074	.07304	.07535	.07764	.07993	.08222	.08450
44	.08677	.08904	.09131	.09357	.09582	.09807	.10032	.10256	.10479	.10702
45	.10925	.11147	.11368	.11589	.11810	.12030	.12249	.12468	.12687	.12905
46	.13123	.13340	.13556	.13773	.13988	.14204	.14419	.14633	.14847	.15060
47	.15273	.15486	.15698	.15910	.16121	.16331	.16542	.16752	.16961	.17170
48	.17379	.17587	.17794	.18002	.18208	.18415	.18621	.18826	.19032	.19236
49	.19441	.19644	.19848	.20051	.20254	.20456	.20658	.20859	.21060	.21261

Table VI *(concluded)*

						500–999				
N	**0**	**1**	**2**	**3**	**4**	**5**	**6**	**7**	**8**	**9**
50	6.21461	.21661	.21860	.22059	.22258	.22456	.22654	.22851	.23048	.23245
51	.23441	.23637	.23832	.24028	.24222	.24417	.24611	.24804	.24998	.25190
52	.25383	.25575	.25767	.25958	.26149	.26340	.26530	.26720	.26910	.27099
53	.27288	.27476	.27664	.27852	.28040	.28227	.28413	.28600	.28786	.28972
54	.29157	.29342	.29527	.29711	.29895	.30079	.30262	.30445	.30628	.30810
55	.30992	.31173	.31355	.31536	.31716	.31897	.32077	.32257	.32436	.32615
56	.32794	.32972	.33150	.33328	.33505	.33683	.33859	.34036	.34212	.34388
57	.34564	.34739	.34914	.35089	.35263	.35437	.35611	.35784	.35957	.36130
58	.36303	.36475	.36647	.36819	.36990	.37161	.37332	.37502	.37673	.37843
59	.38012	.38182	.38351	.38519	.38688	.38856	.39024	.39192	.39359	.39526
60	6.39693	.39850	.40026	.40192	.40357	.40523	.40688	.40853	.41017	.41182
61	.41346	.41510	.41673	.41836	.41999	.42162	.42325	.42487	.42649	.42811
62	.42972	.43133	.43294	.43455	.43615	.43775	.43935	.44095	.44254	.44413
63	.44572	.44731	.44889	.45047	.45205	.45362	.45520	.45677	.45834	.45990
64	.46147	.46303	.46459	.46614	.46770	.46925	.47080	.47235	.47389	.47543
65	.47697	.47851	.48004	.48158	.48311	.48464	.48616	.48768	.48920	.49072
66	.49224	.49375	.49527	.49677	.49828	.49979	.50129	.50279	.50420	.50578
67	.50728	.50877	.51026	.51175	.51323	.51471	.51619	.51767	.51915	.52062
68	.52209	.52356	.52503	.52649	.52796	.52942	.53088	.53233	.53379	.53524
69	.53669	.53814	.53959	.54103	.54247	.54391	.54535	.54679	.54822	.54965
70	6.55108	.55251	.55393	.55536	.55678	.55820	.55962	.56103	.56244	.56386
71	.56526	.56667	.56808	.56948	.57088	.57228	.57368	.57508	.57647	.57786
72	.57925	.58064	.58203	.58341	.58479	.58617	.58755	.58893	.59030	.59167
73	.59304	.59441	.59578	.59715	.59851	.59987	.60123	.60259	.60394	.60530
74	.60665	.60800	.60935	.61070	.61204	.61338	.61473	.61607	.61740	.61874
75	.62007	.62141	.62274	.62407	.62539	.62672	.62804	.62936	.63068	.63200
76	.63332	.63463	.63595	.63726	.63857	.63988	.64118	.64249	.64379	.64509
77	.64639	.64769	.64898	.65028	.65157	.65286	.65415	.65544	.65673	.65801
78	.65929	.66058	.66185	.66313	.66441	.66568	.66696	.66823	.66950	.67077
79	.67203	.67330	.67456	.67582	.67708	.67834	.67960	.68085	.68211	.68336
80	6.68461	.68586	.68711	.68835	.68960	.69084	.69208	.69332	.69456	.69580
81	.69703	.69827	.69950	.70073	.70196	.70319	.70441	.70564	.70686	.70808
82	.70930	.71052	.71174	.71296	.71417	.71538	.71659	.71780	.71901	.72022
83	.72143	.72263	.72383	.72503	.72623	.72743	.72863	.72982	.73102	.73221
84	.73340	.73459	.73578	.73697	.73815	.73934	.74052	.74170	.74288	.74406
85	.74524	.74641	.74759	.74876	.74993	.75110	.75227	.75344	.75460	.75577
86	.75693	.75809	.75926	.76041	.76157	.76273	.76388	.76504	.76619	.76734
87	.76849	.76964	.77079	.77194	.77308	.77422	.77537	.77651	.77765	.77878
88	.77992	.78106	.78219	.78333	.78446	.78559	.78672	.78784	.78897	.79010
89	.79122	.79234	.79347	.79459	.79571	.79682	.79794	.79906	.80017	.80128
90	6.80239	.80351	.80461	.80572	.80683	.80793	.80904	.81014	.81124	.81235
91	.81344	.81454	.81564	.81674	.81783	.81892	.82002	.82111	.82220	.82329
92	.82437	.82546	.82655	.82763	.82871	.82979	.83087	.83195	.83303	.83411
93	.83518	.83626	.83733	.83841	.83948	.84055	.84162	.84268	.84375	.84482
94	.84588	.84694	.84801	.84907	.85013	.85118	.85224	.85330	.85435	.85541
95	.85646	.85751	.85857	.85961	.86066	.86171	.86276	.86380	.86485	.86589
96	.86693	.86797	.86901	.87005	.87109	.87213	.87316	.87420	.87523	.87626
97	.87730	.87833	.87936	.88038	.88141	.88244	.88346	.88449	.88551	.88653
98	.88755	.88857	.88959	.89061	.89163	.89264	.89366	.89467	.89568	.89669
99	.89770	.89871	.89972	.90073	.90174	.90274	.90375	.90475	.90575	.90675

Source: Reprinted with permission from *CRC Standard Mathematical Tables*, 25th edition, pp. 237–244. Copyright 1978, CRC Press, Inc., Boca Raton, Fl.

Index